# The Emergence Paradigm in Quality Management

reflective paradigm

empirical paradigm

emergence paradigm

reference paradigm

Teun Hardjono · Everard van Kemenade

# The Emergence Paradigm in Quality Management

## A Way Towards Radical Innovation

Springer

Teun Hardjono
De Goudse School
Gouda, Zuid-Holland, The Netherlands

Everard van Kemenade
Van Kemenade ACT
Nuenen, The Netherlands

ISBN 978-3-030-58095-7 ISBN 978-3-030-58096-4 (eBook)
https://doi.org/10.1007/978-3-030-58096-4



This Springer imprint is published by the registered company Springer Nature Switzerland AG
The registered company address is: Gewerbestrasse 11, 6330 Cham, Switzerland

*Gesinnungen leben nicht wenn sie keine*
*Gelegenheid haben zu kämpfen.*

*Thomas Mann*

*When they discovered that they were unable*
*to force a Roman Catholic head upon him,*
*just to be sure, they chopped*
*off his Protestant one.*

*Lichtenberg*

# Introduction

*".... the totality is not, as it were, merely the sum of parts, but the totality is entirely different, from the parts.... " this implies that totality differs from the sum of the parts.*

*Aristotle*

*Emergence is a particularly key concept in system theory and the philosophy of science. It often involves the development of complex, organized systems, exhibiting certain properties that cannot be traced back to the properties of their constituent parts.*

*Wikipedia*

With our book, we strive to introduce emergence as a paradigm in quality concepts. To both of us, having been actively involved in the quality management subject in our own way over forty years, the feeling remains that we still do not have a sufficient grip on the subject. We have taken note of the unremitting attention to quality in Japan, which arose in the early fifties. We participated in the quality management boom that occurred in both the United States and Europe in the eighties of the previous century. Teun intensively contributed to the development of the European quality management model, the so-called EFQM Excellence Model (European Foundation for Quality Management) and especially to the Dutch version, the INK (Institute for Dutch Quality) Management Model, as well as to the organization's establishment. Everard made significant contributions to the application of EFQM in higher education. We have personally met almost all the major gurus the quality management field has produced. In the meantime, we have seen the quality system certification based on the ISO 9000 series soar, while much attention remains for programs such as Lean and Six Sigma. Both of our theses address the topic of quality in organizations. While one thesis addresses quality management applied to education, the other is an effort at gaining a stronger grip on organizational development.

Undeniably, the attention given to quality has produced some results. The efforts regarding this topic must have paid off. Otherwise, the focus on quality management would have vanished from the scene long ago. Granted, the hype may have passed its peak. Yet the attention for quality remains and many of the contributions have gained a permanent status within organizational management, with certification perhaps being the most striking.

Obviously, various questions did arise over the course of time: How is it that after some initial enthusiasm the attention to quality of the top of organizations cools down again? Why do we see many references to Robert Pirsig's *Zen and the Art of Motorcycle Maintenance*, yet so little is actually done with it? Why does it seem that attention for quality and innovation are total opposites? Why is attention for quality particularly hard to achieve within the health care sector? And why does health care stick to approaches different from those applied in other sectors? It appears as if the available set of tools is insufficient. (Of course, existing instruments may need improvement, but does quality science itself have a need and a potential for innovation? This is a rhetorical question as far as we are concerned. Particularly, if you define innovation as "a discovery or invention applied which leads to a paradigm shift, whereby the discovery or invention itself is not a linear continuation of what already exists").

In our book we describe the existing streams of thought regarding quality. We indicate why we consider the paradigm concept as the most adequate term within this context. Initially, we will provide a brief historical summary of the thoughts regarding quality and the exploratory quests made by various influential thinkers. In such a way we want to demonstrate that the search for another, complementary, paradigm is not only appropriate, but actually also logical and needed. We called it "the Emergence Paradigm", which also serves as the book's title.

The concept of "emergence" is probably not well-known among a broad public, yet it is an old concept. In fact, it was even described by Aristotle. Perhaps the concept has been somewhat forgotten, due to the Age of Enlightenment. However, we see that it has gained new attention in various places. Nonetheless, in our opinion, the concept has not yet received a broad application in the world of quality management. Maybe this explains some of the disappointment yielded by the attention for quality, aside from notably impressive results. At any rate, an overestimation of other streams of thought transpired, also described here and mentioned as paradigms. In our opinion, the Emergence Paradigm not only richly enhances our thoughts about quality, it also to a degree highlights conditions beneficial to innovation. Hence, the Emergence Paradigm appears to bridge the attention for quality and innovation.

We have allowed ourselves to be inspired by various thinkers and writers: For instance, by the musicologist Rik Spann, who pointed out that making music according to the rules of Jazz is fundamentally different from making music according to the applicable rules for a traditional philharmonic orchestra. Rik Spann also introduced the concept "emergence" to us, when we were looking for a synonym for "pragmatism", as described by Sanne Takema in her lecture "The problem of pragmatism". In this case, Takema made use of Dewey's ideas. The search for an alternative to pragmatism was inspired by the fact that it has obtained a totally different meaning from the one originally intended by Dewey, especially in Dutch language. Rik Spann's reference to Jazz, showed us that beautiful music is not only created by means of a brilliant score under the leadership of a skilled conductor, but that it also "emerges" under the condition that the musicians strictly apply a few simple rules, while individual musicians have the liberty to make their own improvisations. The leader of the band, if there is one "guards the process" but does not exactly determine the band members' performance.[1]

Of course, we also allowed ourselves to be led by our practical experience as organizational advisers. "We are a complex organization and no biscuit factory", is a statement which we as organizational advisers have often heard. Aside from the fact that even a biscuit factory may be fairly complex, it is strange that people nonetheless seem to favor quick solutions. Apparently, the quest for how to turn everything into gold, or finding the holy grail or the philosophers' stone, remains attractive concepts. Moreover, it is a dangerous thought; why else would even Pope Francis in his encyclical Laudato Si warn against simple answers, among other things with the argument:

> *Everything relates to everything else. In this universe, shaped by open and intercommunicating systems, we can discern countless forms of relationship and participation. It cannot be emphasized enough how everything is interconnected. Time and space are not independent of one another, and not even atoms or subatomic particles can be considered in isolation*, said Pope Francis (LS138).

What we want to say is that we allow ourselves to be inspired by what we have seen in the field of systems thinking, theory of complexity, self-organization, and *sense-making*.

Cor van Beuningen (2015), in his thoughts on Laudato Si (LS) writes:

> The concept of integrated ecology, which occupies a central position in LS, is closely linked to systems thinking and the theory of complexity, which have considerably evolved in the past two decades. Integral ecology represents the idea that reality is a complex, systemic whole, which in itself consists of open, living systems. The system components – humans, animals, objects, organizations, subsystems – are interrelated in many ways and can only be understood within their mutual correlation. As they interact they influence and change each other; they co-evolve. In their interaction, they generate new properties (so-called emergent phenomena) with the result that the system's behavior cannot be reduced to that of its components. Through feedback mechanisms, these system properties in turn influence the components and their relations (backward causation) The system establishes a relation logic between the parts, usually self-organization patterns. Complex systems in themselves are a part of complex larger systems. Moreover, between these levels there are also interactions with feedback loops (nestedness). (Beuningen, C van. (2015) Laudato Si voor wetenschap en bestuur. Interne notitie st. Socires. The Hague. www.degoudseschool.nl, theme: Our Common Future).

In our opinion, we find ourselves among good company.

## The Objective

In our book, we will describe thinking about quality as thinking from different paradigms. We have reached four paradigms, whereby the first three have been

[1]See also: Spann (2018).

commonly accepted for a long time. Yet we also demonstrate that a gap exists, allowing us to formulate a fourth paradigm. We indicate why we think that the "paradigm" concept is appropriate and how the four paradigms relate to other theories. We describe the role and the meaning of the context in order to position and prioritize the four paradigms in relation to one another. The book concludes with some afterthoughts and reactions, which various colleagues from our field have suggested to us. We do this, since we intend to enter into a dialogue about how to understand quality management, how to apply it, what may be expected of it, and how it fits within the existing know-how of a sound organization.

Therefore, let us start with a short summary of the evolution in the thinking about quality. In this way, we want to share with the reader our quest for the fourth paradigm. We hope it offers an answer to our questions and at the same time accommodates the quickly changing complex context of organizations existing in the current day and age.

Our book is intended for strategy and policy makers, wanting to establish a link between their conceived plans and the attention for quality. Hence it is also intended for everyone involved with quality in various ways, be it as an adviser or as a quality official. It is not an easy book. So we do not provide seven, ten, twelve, or twenty-five ways for guaranteed success or eternal glory. However, the number four will occur fairly often, because we are convinced that aside from commonly known paradigms within thinking about quality, there is yet a fourth. We will make it plausible that the fourth, Emergence Paradigm, completes the palette for thinking about quality. The latter makes it possible for strategy and policy makers to choose which approach best suits their intentions: For advisors, this entails weighing the various options one against the other. For the quality officer, this enables him to get a clear picture of what instruments he or she will be using in his or her organization and what arguments will have to be put forward to managers in order to gain their support.

We are convinced that it is vitally important to also provide a contribution to working on innovation. It is not without reason that there have been various quests for determining the relationship between quality management and innovation. At the same time, it always resulted in a paradox. Quality management appears to make promises that everything can be brought under control, that no or very few mistakes will be made, and that managers and supervisors will not be surprised by unexpected situations. Whereas innovation not only promises just that, it is even searches it. Innovation searches not only for mistakes, often resulting in innovation, yet particularly searches for surprises. The Emergence Paradigm indicates what the preconditions for reaching the objective.

A summary of our ideas was published in English, in *TQM Journal* of February 2019 (Van Kemenade & Hardjono 2019). Our book further elaborates on this theme.

## De Goudse School

The book is an initiative of De Goudse School (www.degoudseschool.nl), offering a platform for organizational issues, a venue where one may potentially enter into discussion.

A quote of Thomas Mann: *Gesinnungen leben nicht wenn sie keine Gelegenheit haben zu kampfen*, (free translation: Attitudes are not alive when they have no opportunity to fight). Since we do not claim to have a monopoly on wisdom, yet we want to provide a contribution to the dialogue about quality. In our view, the naming of these four paradigms does not result in a prescriptive model. It should be regarded as an effort to give meaning to the ways in which people view quality. In quality science until now, we lacked a consistent manner of viewing, of organization, and of decision-making. We think that what we later will refer to as the Emergence Paradigm, make a significant and refreshing contribution to the field of quality science.

> We do not claim to have a monopoly on wisdom, but we wish to make a contribution to the dialogue about quality.

For this reason, we have submitted our manuscript to Huub Vinkenburg, Henk de Vries, Arend Oosterhoorn, Kees de Vaal, Olaf Fisscher, Rik Spann, Arnold Roozendaal, and Ben van Schijndel and listened to their comments, even though we did not accommodate all of their points. We have given them a chance to share their own ideas or comments as an appendix in our book. Additional comments are always welcome, even though we like to take into account Lichtenberg's warning: *"When they realized that they could not force a Roman Catholic head upon him, just to be sure, they chopped of his Protestant one"*.

Gouda, The Netherlands
Nuenen, The Netherlands
April 2020

Teun Hardjono
Everard van Kemenade

# Contents

# About the Authors

**Em-Prof. Dr.Ing. Teun Hardjono** (1949) began his career in 1976 with the Berenschot organizational consultancy bureau. The first years, he focused on the confection industry and produced a number of international orders. His involvement in quality management as a professional field began when he became the secretary of the Berenschot quality council. They studied the question how to obtain a grip on the quality of a service rendering professional organization. It was a question which previously had not been explicitly studied. Hardjono turned his acquired insights into an advice program. Berenschot was the first organizational bureau that entered the market with this concept. In 1995, Hardjono obtained his Ph.D. at Technical University Eindhoven based on the thesis: getiteld *Ritmiek en Organisatiedynamiek*. From 1998 to 2015, Hardjono had an appointment as a part-time full professor in Quality Management and Certification at the Rotterdam School of Management (RSM), Erasmus University Rotterdam. He supervised a large series of graduation theses and served as the promoting professor of a dozen of Ph.D. students.

As the first Dutchman ever, Hardjono was invited to join the prestigious International Academy for Quality and is the founder and chairman of the Dutch Academy for Quality. He is the spiritual father of the Instituut Nederlandse Kwaliteit giving an impulse to the establishment of MVO Nederland.

Hardjono is a member of the consultative group Philosophy of Life and Enterprise of the VNO-NCW and has served as a municipal council member for the CDA (Christian Democratic Party) in Gouda. He is one of the founders of DeGoudseSchool.

**Dr. Everard van Kemenade** (1953) began his career in 1972, as a professor of Dutch language in Higher Vocational Education. Quite soon, he was invited to consult on educational innovation projects. Later, he was invited to lead such projects. He took additional training courses as an educational consultant from Twente University. Van Kemenade discovered his first passion when invited to prepare training courses for visitation and later for accreditation. During the rest of his academic career he remained involved in quality science. He discovered his second passion when invited by Wageningen University in 2004 to perform a quality project for a university in Vietnam. Since then, working on quality in other cultures has become the core of his work. In 2009, Van Kemenade obtained his Ph.D. from the Rotterdam School of Management, Erasmus University, based on his study into the question of how the professionals in higher professional education in The Netherlands and Belgium experience accreditation. Since then he has published various articles in magazines in the field of quality such as: Synaps, Sigma, TH&MA, Quality in Higher Education, Quality Progress, International Journal for Quality and Reliability Management, and the TQM Journal. He is a member of the Dutch Academy for Quality. Since 2017, he has been primarily involved in the field of quality in the Caribbean, Africa, and the Middle East, representing his own consultancy, Van Kemenade Audit Coaching and Training (see: www.vankemenade-act.nl). He also works three days a week as a senior lecturer and researcher in the Master Integrated Care Design of Utrecht University of Applied Sciences.

# Chapter 1
# The Evolution of Thinking About Quality

*Whoever does not know the past will not have a grip on the future.*

Golo Mann.

The attention for quality has existed as long as products have been manufactured. Even in the Stone Age (in Europe going as far back as 4000 BC) products were compared to a previously determined norm. An early proof of the existence of quality science dates back from 3000 BC. Drawings discovered on Egyptian pyramid walls exhibit an extensive quality control and quality inspection system which was properly utilized in their construction projects. The Hammurabi Code, dating back from the Babylonian era (1728–1686 BC), imposes heavy punishments on contractors, when the quality of their work was so defective that others might get hurt, or incur other types of damage.

## 1.1 From the First Quality Certificate

The first quality certificate known to us dates back from approximately 429 years before Christ: It is a stone tablet with the following inscription: "*Regarded the gold ring set with an emerald, we guarantee that the emerald will not fall out of the setting for a period of twenty years. If the emerald should fall out of the ring prior to that time, then we will issue a compensation of ten mana of silver to Bel-nadin-shumu*".[1]

The ancient Greeks contemplated quality. Plato's *Protagoras* from that era is a dialogue between Socrates (philosopher from Athens) and Protagoras (a Sophist from Abdera). Socrates requests Protagoras to demonstrate how you can learn to be excellent in citizenship. Their dialogue proves how difficult it is to properly define quality. Juran (1995) describes among other examples developments in Ancient China, Ancient Israel, and Early India. In these early days, when business transactions

[1]Retrieved from the booklet "Van Plato tot Pluto" Instituut Nederlandse Kwaliteit, 1996.



T. Hardjono and E. van Kemenade, *The Emergence Paradigm in Quality Management*, https://doi.org/10.1007/978-3-030-58096-4_1

took place directly between the individual producer (the blacksmith, the tailor, the farmer, etc.) on the one hand and the individual user on the other, perhaps the needs of the users were easily ascertainable. Yet, it is questionable whether what they wanted was "quality". During the Middle Ages, the trade guilds evolved, as the number of suppliers increased and the line from producer to consumer became more complex. In fact, the trade guilds guarded the quality for a particular group of tradesmen, the professionals of their day. Guilds imposed quality norms for production, use of materials, end products describing specific ideas on how know-how was acquired and how it was to be conveyed. The key concept in this quality control approach was the maker's craftsmanship. Product quality was checked by institutions established particularly for this purpose.

Shewhart (1931) was the first modern expert wrestling with the concept *quality*. He distinguished two aspects: the objective aspect (an objective reality independent of the existence of men) and a subjective aspect (what we think, feel, or sense as a result of the objective reality). It was not until after the Second World War that the concept of quality was used on a broad scale. In our day and age, the direct relationship between producer and consumer has often given way to an entire supply chain (consisting of producer, wholesaler, retailer, and consumer). With the exception of those situations (a growing phenomenon) in which products are made "to order" (i.e. according to the customer's specification)—the producer determines which products he manufactures how. He may do this based on information gleaned from, for example, market research. Indirect relationships between producer and consumer led to a situation in which people attribute an absolute meaning to the concept "quality".

## 1.2 EOQ and EFQM

In 1946, the American Society for Quality (ASQ) was founded in the United States, driven by the attention for quality which existed prior to World War II. Somewhat later, in 1957, the European Organization for Quality (EOQ) was established in Europe. From its beginning, the ASQ was an organization for quality experts, divided into sections and divisions, whereby each section focused on a particular product category (automotive, food, aerospace, etc.). The EOQ was an umbrella organization of which originally each European nation provided a delegate from a national quality organization. For the Netherlands this initially was the so-called KwaliteitsDienst voor de Industrie (KDI). Also, the EOQ was subdivided into sections intended for a particular group of producers. ASQ and EOQ formed one way of thinking about quality.

In the eighties of the previous century a new organization had to be established to promote quality, particularly to policy makers and entrepreneurs. The government of the United Kingdom gave more attention to quality under the political leadership of his kindred spirit, Margaret Thatcher. British Standard BS-5750, the result of NATO's Allied Quality Assurance Publication, played a compelling role in exports and imports to and from the United Kingdom. Shortly thereafter, the BS-5750, together with the German DIN 8563, the Dutch NEN and NPR (Netherlands Practical Guideline) 2646, formed the basis of what we now know as the ISO 9000 series. Many years preceded the creation of an ISO standard. Most strikingly, simultaneously the Malcolm Baldrige Award Criteria (MBAC) were gaining traction. These criteria were defined by a law passed under the Reagan administration. Naturally, people in Europe wanted to invent their "own wheel" with the European Foundation for Quality Management (EFQM). The signing of the declaration of intent occurred in 1988, resulting in its formal founding in 1989. In 1991, Teun Hardjono, together with his Berenschot colleague, Aalco van der Veen, was ordered by the Dutch Ministry of Economic Affairs, to conjure up some advice concerning how the existing quality program might receive a further impulse. It was particularly a mediation attempt to align a great number of competing organizations, each with attention for quality in their objectives. Based on their advice, the Institute for Netherlands Quality (INK) was established. Soon after, INK launched the INK management model, merging the EFQM model with the Berenschot Generation model. ISO, MBAC, and EFQM formed another way of thinking about quality.

In his inaugural speech "Tacking between liberals and purists", Hardjono (1999) already referred to the above-mentioned schism in the world of quality management. He referred to the EFQM c.s. as the world of liberals and the traditional supporters of the ASQ (and EOQ) as the world of purists. In the meantime, it has become clear to us that dividing the world into *liberals* and *purists* is a division into two different paradigms. Thoughts in the world of EOQ and ASQ are dominated by the statement: *measuring is knowing*. Huub Vinkenburg labels it as *empirical school*. We follow him in this, although we prefer to substitute the concept *school* by *paradigm*. Therefore, we refer to this as the *Empirical Paradigm*. It fits with the so-called *Anglo-Saxon model*, focusing on shareholders.

At the base of initiatives, such as the ISO 9000 series, the Malcom Baldrige National Quality Award (MBNQA) criteria, and the European Business Excellence Model (EFQM), lies the introduction of a frame of reference, a model. This model, a construct, is a means by which organizations—in and of themselves constructs—are tested. Measuring absolutely, even though this is implied, is no longer possible. One will have to measure with a whether or not inter-subjective judgment. This needs to be handled by those familiar with the model. They are led by the criteria/norms (MBNQA and ISO 9000) of areas of address (EFQM model and INK management model). Vinkenburg for this reason, by recommendation of Olaf Fisscher, labeled this the *normative school*.

As Hardjono points out in his inaugural speech, it is a different approach rather than a stream of thought or an orientation. Yet, at a greater depth, supporters of the notion that only empirical facts provide scientific knowledge, the so-called positivism, certainly reject the ideas behind the *normative school*. Interestingly enough, the utilized models are a collection of opinions that cannot be falsified and the applied measuring method due to their intrinsic subjectivity unrepeatable. In what we labeled the *Empirical Paradigm*, the principles of *positivism* are followed as a guideline. *Normative* reminds us too much of what is applicable within the Empirical Paradigm, addressing only one even though entailing more aspects. Particularly since this concerns a different mentality and attitude, we prefer the concept *Reference Paradigm.* In the *Reference Paradigm* a quality evaluation is based on comparison within a frame of reference, with the norm or model coming about by mutual consultation. It belongs to the scientific school of thought referred to as: *social constructivism.* Social constructivism is the theory that our experience of the world is (partially) constructed through the social processes relevant to the society in which we live. This explains why there are various models, and why there are significant differences between the American Malcom Baldrige Award Criteria model and the European Foundation Quality Management model. The American model, for instance, treats the employees as the most important *resources* of an organization and considers leadership, as well as the reward for it, to be linked to one individual. The American model is also based on the fact that there is only one stakeholder, usually the shareholder, who is the stakeholder. The European model, on the contrary, regards the employees as important building blocks of the organization and does not merely treat them as resources, with leadership as an interactive process in which both parties depend on one another. Therefore, both parties are responsible for each other's behavior. The most characteristic aspect of the EFQM model is that it explicitly bases itself on a multi-stakeholder approach. It concerns a balancing of interests among all stakeholders, the so-called *Rhineland model.*

However, with the rise of the Reference Paradigm, the Empirical Paradigm did not disappear. For various reasons, mutual misunderstanding and irritation caused the attempts to merge EOQ with EFQM to fail miserably. The failure was also caused by the incompatibility of the tempers of the leading figures of the era, yet particularly by the different perspectives. In his book, *The Structure of Scientific Revolutions* (1962), Thomas Kunhn argues that it is impossible to merge the two paradigms. His definition of *paradigm* allows us to prefer this word over the concept of *schools.* We will further address the issue in a later chapter.

## 1.3 Health Care Provides a Third Paradigm

Both above-mentioned paradigms have been utilized in health care. Health care is partly providing a "cure" with the striving for "evidence based medicine". The Empirical Paradigm is most useful for thinking about quality in the health care context. Yet, models and norms are also popular; just consider the accreditation standards of the Joint Commission International and the EFQM Excellence Model from the Reference Paradigm.

For a major part, health care is *care* based, closely akin to providing services with questions about life and death. This implies that there is room for, or perhaps even a need for another paradigm. The ability to determine the quality aspects of care, the possibility to form an opinion about it, and to look for control and improvement always resonates in the provider's as well as in the consumer's role. Both the consumer and provider are individuals with all of their peculiarities and uniqueness. Pirsig (1976) beautifully describes this in *The Art of Motorcycle Maintenance.* His book is particularly philosophic in nature, bringing up questions such as: "But what do you think of it yourself?", "What do you really think of it?", "Have you sufficiently done your best?", and "What image of man do you take as a starting point?". Hence, Vinkenburg recognizes a third school of thought: the *reflective school.* Even the reflective school we consider to be more than a manner of acting. So, we refer to this approach as the *Reflective Paradigm.* Within the Reflection Paradigm, participants come to a conclusion based on mutual consultation/on-going discussion. It relates to the scientific philosofy of inter subjectivism.

## 1.4 A Fourth Paradigm

However, there are many paradoxes and dilemmas for health care in particular. Complete with an ever-shifting horizon, health care accommodates not only new technological challenges but also addresses moral and ethical issues. For that purpose, the three existing paradigms are no longer sufficient. One should also be able to think and act from a new paradigm in which the context plays an important role. The three paradigms no longer suffice to control our actions. This prompts us to think about a fourth paradigm, which particularly applies in situations with a rapidly changing context.

It is clear that the three existing paradigms, for example, are unable to sufficiently explain the history of Japan's success by giving attention to quality. Whereas in the Western world efforts were made toward ISO, Malcolm Baldrige, and EFQM. From 1951, people in Japan simply continued with their own quality movement in the form of the Deming Prize. This prize was first awarded to Fuji. Surprisingly enough, the Japanese program was based on the ideas of quality management gurus Deming and Juran and initiated by General McArthur, representing the United States while taking charge of Japan's restoration. For a long time, experts assumed that the ideas regarding the Japanese Deming Prize, the opinions behind the combination of the Malcolm Baldrige National Quality Award (MBNQA), the European Business Excellence model along with their derivatives, such as the Australian Quality Framework and even the ISO 9000-series, might best be combined under a single heading, particularly in their most recent form. As such, the Deming Prize was incorporated within the Reference Paradigm. Based on our current insights, we believe that such an incorporation is incorrect. Since we gained more insight in a fourth paradigm, what we refer to as the Emergence Paradigm, we have reached the conclusion that the Deming Prize does not belong to the same paradigm as the MBNQA. JUSE, the Union of Japanese Scientists and Engineers, performs a different form of research for its Deming Prize and related NIPPON Award. Again, the most striking aspect is the absence of a reference model, as seen in MBNQA and EFQM. The successful approach of the Japanese cannot be explained on the basis of the old paradigms.

The Japanese Union of Scientist and Engineers, the manager of the Deming Prize, states:

> "The Application Guide for The Deming Prize The Deming Grand Prize 2020" conformance to a model provided by the Deming Prize Committee. Rather, the applicants are expected to understand their current situation, establish their own themes and objectives and improve and transform themselves organization-wide".

In his thesis *Trusting Associations* (2019), Peter Noordhoek describes in great detail two cases related to introducing an association and how to gain control over its members. He describes the different approaches and observes some results, yet these are not fully satisfactory. The approaches, which Noordhoek describes, show major similarities with the above-mentioned three paradigms. He points out another approach that evolved over time. It is the one characterized by pressure from the environment and the willingness to reach a consensus of what is appreciated as *quality* and how to find new ways. Apparently, in modern times with their fast-emerging technological changes, shifting geo-political relations, cross-border issues, and the call for innovations, it is high time to explore a fourth paradigm. We think that we have discovered the fourth paradigm in the Emergence Paradigm. The word *emergence* has not (yet) been adopted in daily common vocabulary, however, what it refers to is definitely not new. It may be used more or less as a synonym for what Dewey by the end of the nineteenth century meant by *pragmatic.* Yet, as we mentioned in our introduction, the word *pragmatic* in the Dutch language received a meaning that we definitely do not wish to link with what we referred to as *emergence.* In Chap. 8, we will be dealing with the concept emergence in greater detail. For now, we would like to refer to the Dutch television broadcasting organization VPRO's series "The mind of the universe" by Robert Dijkgraaf, Director and Leon Levy Professor at the Institute for Advanced Study in Princeton, New Jersey, and President of the InterAcademy Partnership, the global network of academies of science and health, highlighting leading scientists which he interviews often using the expression *emergence.* In which case it stands for more or less evolving insight and for being open to new ideas and evidence supplied.

In *Zen and the Art of Motorcycle Maintenance*, Pirsig (1976) tries to explain what is possible in quality. He does not provide more details, even though, we think it neatly fits into the Reflection Paradigm. In his second book *Lila*, Pirsig tries again and clearly exhibits yet another view on quality. For those who have ever heard the two greatest gurus in quality science, Edward Deming (1900–1993) and Juran (1904–2008), talk about each other, you must have noticed that they had great respect for each other. Yet, they also fundamentally disagreed. Surprisingly, both men have had an indelible effect on the way in which Japan approached and still approaches quality. With our current knowledge we can say that each of them thought from a different paradigm. Juran thought from the Empirical Paradigm and Deming from the Emergence Paradigm. Later, we will discuss this in more detail. The difference of opinion between Deming and Juran is also the reason that we are convinced that Lean, an approach of continuous improvement derived from Japan based on Deming's ideas, may not be mentioned in the same sentence with Six Sigma, a

**Table 1.1** The four paradigms

| Paradigm | Empirical | Reference | Reflection | Emergence |
|---|---|---|---|---|
| Judgment | Objective | Inter-objective | Subjective | Inter-subjective several answers |
| Judgment | Definitive | No absolute yes/no | Eventually there is a judgment, yet it changes over time | Tentative, the truth is debatable |
| Analyst | Measures based on a predetermined standard | Based on knowledge, a third party may give a more broadly accepted judgment | Judgment only valid for the observer | Viewing in different ways |
| Values | Predetermined in a standard measuring method | Anchored in a model | Values are also debatable | Open |

continuous improvement approach from the United States based on Juran's ideas. Nonetheless, the Six Sigma approach shows many similarities with the accelerated education/excellerate training which the consultancy firm Berenschot developed and applied in World War II and afterwards. Around 1950, Berenschot also sold this approach in the United States, while Juran acted as a mentor to the consultants sent out from the Netherlands.

The fourth paradigm is not only a supplement to the ways of thinking about quality. It also helps to explain historic developments and directs our ways to the future. It does not replace the other three paradigms. We also describe the other three paradigms. They remain just as relevant and useful as ever. Indeed, it is a matter of choice which paradigm you should select to apply. The four paradigms demand *epistemic fluency*: that is, the capacity to understand, switch between and combine different kinds of knowledge and different ways of knowing about the world (Markauskaite and Goodyear, 2016).

We use the concept *paradigm* and we will provide the arguments why we prefer this concept instead of schools, phases, streams, philosophies, and the like. In order to make it acceptable why there has to be a fourth paradigm, we will base ourselves on existing theories. These theories not only prove that there is a fourth paradigm but also underscore its characteristics. In Table 1.1 we provide a diagram of the four paradigms as presented and defended in this book.

## References

Hardjono, T. W. (1999). *Laveren tussen rekkelen en preciezen*. Oration Faculty of Business Administration at the Erasmus University Rotterdam. ISBN 90-56773127.

Juran, J. M. (1995). *A history of managing for quality: The evolution, trends and future directions of managing for quality*. Milwaukee, Wisconsin: ASQC Quality Press.

Kuhn, T. S. (1962). *The structure of scientific revolutions.* Chicago/London: The University of Chicago Press. ISBN 0226458083.

Markauskaite, L., & Goodyear, P. (2016). *Epistemic fluency and professional education: Innovation, knowledgable action and actionable knowledge.* Dordrecht: Springer.

Noordhoek, D. P. (2019). *Trusting associations: A surgent approach to quality improvement in associations*, Ph.D. study, Tilburg University.

Pirsig, R. M. (1976). Zen and the Art of Motorcycle Maintenance. Een onderzoek naar waarden. Amsterdam: Ooievaar. ISBN 9057132176.

Shewhart, W. (1931). *Economic control of manufactured product.* New York: D. Van Nosttrad Co., Inc. Reprinted in Sower, V., Motwani, V., Savoie, M. (1995). *Classic readings in operations management* (pp. 191–214). Ft. Worth, Texas: The Dryden Press.

# Chapter 2
# Quality Defined and the Values Concept

*It is impossible to express in words what we are together, the quality represented by our relationship.*

In this chapter we will discuss in more detail the definitions and descriptions of the quality concept. We will need these at a later stage to compare and contrast the differences in thought and the differences in quality paradigms.

## 2.1 According to the Dictionary

The word *quality* stems from the Latin *qualitas,* meaning property, character, or essence. ISO 9000:2005 gives the following definition: *"The degree to which a set of inherent characteristics fulfills requirements".* The Dutch Van Dale dictionary (www.vandale.nl) provides the following definitions of quality:

quality *(the; f;* plural:*qualities)*

1. measure in which something is good; condition, capability, nature: *The quality of this fabric*
2. function: *In his quality as mayor*
3. good capability: *Someone with qualities* skill; *quality products.*

Here, we are concerned with the first description. Van Dale's definition mentions *a condition, capability,* or *nature* but also the *measure in which something is good.*

This gives us two definitions:

*Quality is the capability, nature of something (the how and what of an entity).*

But apparently also:

*Quality is the measure in which something is good.*

The two definitions represent two entirely different opinions. The first more or less states that the quality is objectively measurable and, therefore, it fits well into the



T. Hardjono and E. van Kemenade, *The Emergence Paradigm in Quality Management*, https://doi.org/10.1007/978-3-030-58096-4_2

Empirical Paradigm. The second definition contains a (subjective) value judgment. An explanation of the quality concept that fits both the Reference Paradigm as well as the Reflective Paradigm. It is a view borne out by everyday speech. In everyday speech *quality* stands for having or lacking good or special properties.

## 2.2 A Transcendent Approach

Define quality as the value attributed by stakeholders to *an entity*, so it demands something, an object to be judged. The capability of what? Garvin (1984) has provided some very useful insights in this matter. He distinguishes five approaches: The product-oriented, the process-oriented, the user-oriented, the value-oriented, and the transcendent approach. Let us start with the last approach.

Garvin's *transcendent approach* is beautifully illustrated when Henk Bobbink (2010) provides the quality of his relationship with his wife as an example:

> "It is impossible to express in words what we are together, the quality represented by our relationship. And this is 'ousia', a metaphysic quality concept". And also: "Ousia is a thing in and of itself". Quality manifests itself in different capabilities or schools, but it never allows a description in its fullness".

Quality is recognizable by experience. According to some, quality may be determined absolutely and objectively irrespective of time and place and irrespective of what a particular stakeholder thinks of it at a particular moment. Quality is of an inherent excellence. Quality is that which is unquestionably the best. People talked about different qualities to describe the *type* of product as well as (particularly in the world of advertising) the different qualities of *top quality* products. When cars are discussed, the Rolls Royce is often mentioned. This suggests that quality is a property of the product, irrespective of the purpose and specific wishes of the user. Reeves and Bednar (1994) also distinguished *quality is excellence* as one of the quality definitions. Quoting Pirsig, they state: *"Quality is neither mind nor matter, but a third entity independent of the two.... even though Quality cannot be defined, you know what it is" (o.c. 185)*. According to him, you can recognize quality by experience without being able to exactly define it. In art, the transcendent approach to quality is emphatically present. In *The Timeless Way of Building* architect Christopher Alexander refers to this as:

> quality without a name. "In the same way that we know a good room when we use one, but cannot define exactly what makes it good, we can name the attributes of quality, but cannot define quality itself."

Sower et al. (1995) explain how important the transcendent approach to quality is to advance beyond continuous improvement and to be able to achieve real breakthroughs. They base their views on Plato's four forms of awareness (Table 2.1).

> "Understanding Plato's hierarchy leads to the unavoidable realization that radical improvements (Juran's "breakthroughs", 1970; Barker's "paradigm shifts", 1190; Khalil's "radical

**Table 2.1** Plato's four forms of awareness

| Objects of awareness | Forms of awareness |
|---|---|
| The "good" itself (transcendental quality) | Insight (noesis) |
| Mathematical structures | Understanding (dianoia) |
| Concrete things | Perceptual belief (pistis) |
| Images | Imaging (eikasia) |

innovations", 2000) are only derived from insight—the highest level of Plato's hierarchy – which is most closely related to transcendent quality".

## 2.3 More About Pirsig

In the introduction of our book, we already mentioned that when discussing quality, Robert Pirsig's name is often quoted. He wrote two novels in which the question regarding quality is the key issue: *Zen and the Art of Motorcycle Maintenance* (1974) and Pirsig (1991).

The protagonist of the first book wrestles with the quality concept.

> "Quality….. you know what it is, yet you don't know what it is. But that's self-contradictory. Some things are better than others, that is, they have more quality. But when you try to say what the quality is, apart from the things that have it, all goes poof!"….. (o.c., p. 163). "So round and round you go, spinning mental wheels, and nowhere finding any place to get traction. What the hell is quality? What is it? And what is good, Phaedrus? And what is not good? - need we ask anyone to tell us these things?" (o.c., p. 164). But: "Quality is neither mind nor matter, but a third entity independent of the two…. even though Quality cannot be defined, you know what it is" (o.c., p. 185).

Pirsig is often quoted when authors want to emphasize that *THE* quality does not exist, and that quality cannot be defined. It relates to something that *happens*, that *occurs*.

> "Pirsig provides no ready definition of quality but refers to the meaning of it by means of events in the relationships and interactions between people". And: "Applied to the quality of service, which deals with the relationships and interactions between individuals, the 'quality' concept does not refer to an object or to a characteristic of it, but to an event or encounter. The improvement of services relates to a series of encounters, a continuing interaction, in which changes for good may happen of which the parties involved are aware. Quality is not a property of an object, is not an object in itself, but an event happening at the moment when a subject is confronted with an object" (Vinkenburg 1974).

Also Kelchtermans (2003) states, where the service of education is concerned, good professorship maybe more related to *being someone* (for the students) than *knowing something and being able to do something*. It concerns something that *happens*, that *occurs*, irrespective of the objectives, intentions, or deliberate efforts and that is, therefore, separate from what may be planned or made. Harvey and Green (1993) discuss quality in higher education, mentioning *quality is transformation*.

With this they emphasize the importance of the event experienced by the student. The student, who with added value leaves the educational institution as an alumnus. "*The transformative view of quality is rooted in the notion of 'qualitative change', a fundamental change of form. Ice is transformed into water and eventually steam if it experiences an increase in temperature.*" We do not share these opinions. We do not find the reasoning logical, for an event may just as well be an object, the properties or characteristics of which may be defined.

Pirsig's observation is unmistakably, philosophical, reflective in nature, in which one wants to avoid exact measuring which is so characteristic of the Empirical Paradigm and one certainly wants to avoid the prescriptive Reference Paradigm. They leave very little room for doubt and attempt to define the case. Yet, they do not want to analyze it in a holistic way. Pirsig brings Vinkenburg (2007) to the importance of reflexive thinking. He states that Pirsig also

> "…describes a battle between the rationalistic and hermeneutic science concept, which he refers to as the classical (rational, objective, order-providing) and romantic (relational, subjective, inspiring)". We must not avoid the confrontation between these two ways in which people deal with their own reality. Pirsig demonstrates "how fruitful it may be to allow the development of concepts (thinking and discussing quality, Zen) to run parallel in order to gain experience (being together with his son, doing the maintenance of the motorcycle)" (Vinkenburg 2007).

Vinkenburg sees this as a plea for regular reflection. In this respect, he follows Harteloh (2000), whose entire thesis is devoted to the quality concept in health care. Based on Pirsig, he reaches the conclusion that the quality concept does not *exist*, but that the quality concept simply *originates*, and such—in his opinion—evolves by means of a reflexive process (plan, do, check, act). Other forms of battle in the book are those between vision and action, theory and practice, between part and totality. Even in this respect it all depends on finding the right combination. At the time, Harteloh had not heard of the Emergence Paradigm. If he had, then Harteloh might have typified his view as such.

According to us, limiting the thoughts of Pirsig to the reflective paradigm is wrong. Particularly, when his second publication, Lila, is taken into consideration. Pirsig's name is quoted when authors want to emphasize that quality cannot be defined. So, the protagonist of the first book wrestles with the quality concept. However, Pirsig's second book, *Lila*, is much more concrete. Pirsig (1991) says: "*Quality. Quality was value. They were the same thing*" (o.c., p. 66/67). Pirsig distinguishes the dynamic from static quality. And within the static quality, he prescribes the following value hierarchy:

1. Anorganic values
2. Organic values
3. Social values
4. Intellectual values.

In his doctoral thesis, *Rhythm and Organizational Dynamics*, Hardjono has adopted these values, even though he prefers the *powers* concept, especially in the sense of *potency*. The doctoral thesis discusses how organizations evolve over time, while emphasizing the continuous enlargement always of a different kind of power, while the other three powers provide support in this context. Later, we will deal with the outcome of his research, the so-called *four-phase model*, the scientific plausibility he eventually demonstrates.

But, according to Pirsig, quality is not only static and measurable at a certain moment; it is also dynamic. Quality is time and context-dependent. Pirsig refers to Victorian values as an example. **This vision of quality no longer fits the Empirical Paradigm, Reference Paradigm or Reflective Paradigm. This vision of quality fits within the Emergence Paradigm, as we will explain below in more detail.**

## 2.4 Product or Service

We mentioned that Garvin (1984) distinguishes a number of approaches to quality. We already mentioned the transcendental approach. Actually, the first approach he mentions is the *product-based approach* with product quality as the key element.[1] It concerns products (or services) meeting the specifications, previously determined standards, exact variables. "*Differences in quality amount to differences in the quantity of some desired ingredient or attribute*".

Crosby (1979) talks about *quality* in its meaning of *conformance to requirements*. The influence of statistics is dominantly present, particularly in this approach. Crosby was convinced that the cost of production may be reduced through quality improvement. The fastest way to achieve quality is by prevention. The performance norm should be: zero defects. The measure for your quality is the price of non-conformance (i.e. not meeting the requirements) and you should statistically track all such instances.

Reeves and Bednar (1994) mention as one of their quality definitions: *Conformance to specifications* (according to Gilmore, 1974; Levitt, 1972). Often, instead of specification or requirement the terms *norm* or *standard* are utilized. The definition of standard according to ISO is: "*document, established by consensus and approved by a recognized body, that provides, for common and repeated use, rules, guidelines or characteristics for activities or their results, aimed at the achievement of the optimum degree of order in a given context*" (ISO/IEC Guide 2:2004). The clearer and the more measurable the norms, the better we get a grip on the question whether an object meets the norms, which determines its quality. This definition of *quality* matches with the Empirical Paradigm, which we will be describing in more detail below.

The quality of a product is translated by Garvin (1987) in the following eight features:

1. The product's performance (for a car: fuel consumption, safety, acceleration etc.);

[1]In Garvin's description of this approach, the evaluation of the product is only quantitatively described. We will deal with this later in paragraph 2.2.

2. The secondary product characteristics (for a car: CD player, navigation system, paint);
3. Reliability (how long will the car function without any problems);
4. Conformity (to what extent does the car meet the specifications);
5. Durability (how long will the car be used before it incurs major expenses);
6. Ease of maintenance (how easy and affordable are repairs and maintenance);
7. Aesthetic value (how beautiful, pleasant, comfortable seating, or even the scent of the vehicle; the color, the type of upholstery);
8. Experienced quality (brand name, image, reputation, advertising).

The entity may also be a service. Zoest (2001) mentions one of the differences between production firms and service providing organizations within the non-profit sector: no return shipments are possible for a rendered service. "*Customers or participants utilizing institutions providing services, may never return the service. It is possible to file a complaint, yet the customer must be rather assertive to do so. This may not be feasible for all customers in the sector of education, health care and wellness. In fact, customers within service providing organizations only know in hindsight what they received or bought*".

The quality of a service is translated by Garvin in the following eight features:

1. Timeliness (for banking business or paying an invoice via Internet: How quickly will they assist me?)
2. Completeness (Will I get all information that I need?)
3. Respect (How does the bank employee treat me? Or how am I treated by the software?)
4. Consistency (Do I have to be at the same desk for a similar question or do I have to go through various procedures for similar questions?)
5. Accessibility (What is the distance between my house and the bank, or how much banking business can I do via Internet from home?)
6. Ease (How many forms do I have to fill out or how many passwords do I have to type in before I have reached my bank statements?)
7. Accuracy (To what degree are the facts of my Internet account correct?)
8. Flexibility of the response (How long will it take before I receive an answer to my question?)

In the context of the provided service and particularly of interaction-related service provision, the concept *quality* does not refer to an object or aspect, but to an event or encounter affecting the people involved, according to Vinkenburg (1995). Later, we will discuss this in more detail.

## 2.5 Process

In addition to the product/service, we may define the (production) process as an entity. Aside from a product-oriented approach, Garvin (1984) provides this as a second approach to quality, the process-oriented approach (see Table 2.2).

**Table 2.2** Product quality versus process quality

| | Product quality | Process quality |
|---|---|---|
| Type of requirements | Product requirements | Process requirements |
| Type of quality care | Quality inspection | Quality control |
| Type of control | Exit control | Control circuits in the process |

The process approach is originally derived from operations management. Crosby's ideas about *conformance requirements* may also be translated into process requirements, the requirements of the input and output of the process steps. Garvin calls this his third approach: the *manufacturing-based approach.* It concerns *making it right the first time* (Crosby 1979). The idea is that we do not need to check each product if we truly control the process by which the products are made. Traditionally, a distinction is made between primary and secondary processes. Taguchi (1992) emphasizes that not only the manufacturing or the production processes are important for determining product quality. He focused on the integration of product specifications, already in the design process. Processes are often distinguished in primary and secondary or supporting processes. Together, these constitute a system. The quality of processes with Six Sigma is converted into the following five facets of the so-called *DMAIC-cyclus*:

1. Define: Definition of the process
2. Measure: Determining and measuring of process quality at this moment
3. Analysis: Determining of the relationship between the process characteristics and the result
4. Improve: The improvement of the process performance by creating an optimal configuration of the process
5. Control: Control of the achieved situation and the level of process quality.

(Source: Hardjono and Bakker (2002), p. 226)

From the attention for (the improvement of) quality in the ways hereby described, a definition may be deducible particularly aimed at the activities, or the care that should lead to (an improved) quality.

## 2.6 Profession

Aside from the product/service and the process, the profession may be named as an entity. Stevens et al. (1999) mention management of the profession as core to the INK management model in addition to management of processes (see Fig. 2.1).

*"What makes organizations unique, is the way in which they have found an approach to fulfill the customer's needs, thanks to a sound process management. On the one hand, this may be accomplished by a strong* control *of the process-based*

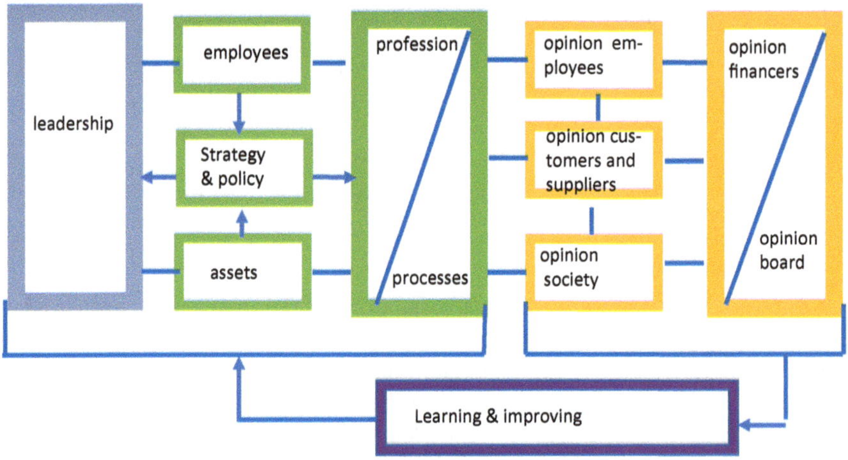


**Fig. 2.1** Profession in the INK (Dutch) version of the EFQM excellence model according to Stevens et al. (1999)

*aspects of the processes. These aspects may be procedurally arranged and formalized. Moreover, organizations distinguish the professional aspects of the processes, whereby the stacked experience of various professional views of reality must have led to the processes in the current form" (p. 216).* Freidson (2001) mentions the importance of the profession (professionalism) *the third logic* after the primate of the manager (managerialism) and the consumer (consumerism).

When in many professions, particularly in the services sector, the interaction between professional and stakeholder is crucial, then the preservation and performance of professionalism are also crucial. This professional factor, moreover, also has its own view on the quality of product, service, process of organization, particularly from professional craftsmanship and professional maturity. The professional may then decide against the wish of the customer or manager, against the procedures and then choose for his own vision of quality and the related actions (and to account for it afterwards).

It is also defensible that this is not a separate object, but a differentiation of the process approach.

## 2.7 The Organization/the System

The European Foundation for Quality Management model and the INK management model distinguish nine areas of attention in a company which together determine its quality. We already showed these in Fig. 2.1: Leadership, Strategy and Policy, Means and Partners (assets), Employees, Processes, opinions of the employees, customer and suppliers, society, and end results. The INK management model and the

EFQM Excellence Model highlight aspects, *areas to address* whereby the judgment processes of an organization must be checked against all of these nine areas. The INK management model and the EFQM Excellence Model fit the Reference Paradigm. In a holistic way, through mutual correlation, these two models predetermine choice aspects essential for the quality of an organization.

*From product, service, and profession to system chain and society*

When we examine the developments in quality thinking, you may observe a shift from a product-oriented approach via a more process-oriented or system-oriented approach to an approach driving the quality of the entire organization. The organization may also be an object of quality, for example when the company wants to obtain an ISO 9001:2015 certificate or to compete for the European Quality Award or the Netherlands Quality Prize. This shift was identified in the developmental phases of the INK management model.

The INK management model originally distinguished five phases of development, derived from the Berenschot Generation Model.[2] (Hardjono and Hess 1993). Each individual organization may have their own interpretation of the nine areas of attention derived from the European EFQM model. In fact, the INK management model specifically mentions the organization with a product-oriented quality focus (or in case of a service: activity-oriented). This is the first phase of development of an organization toward excellence. The INK management model calls an organization process-oriented when its focus is on process quality, the second developmental phase. The third developmental phase is system-oriented. The organization realizes that quality is not only delivered by control of the primary process but also by the supporting processes, and particularly by the correlation between primary and secondary processes, with properly functioning interfaces. In phase four of the INK management model the organization targets the partners in the chain. The INK management model also defines a fifth developmental phase: learning and excelling. This fifth phase reaches the highest degree of quality awareness.

We have always objected to this conclusion. Not only would society-oriented be more logical; the scope of attention "grows", but it does not endorse a hierarchy. The focus on corporate social responsibility (CSR) may, therefore, be regarded as a natural consequence of working on quality. Hardjono intentionally advised the Netherlands Ministry of Economic Affairs at the time to take INK as an example of how to give more attention to corporate social responsibility in the Netherlands. Incidentally, it could be noted for non-profit organizations that the "reverse" way of paying more attention is also true: a better society (often the objective of a non-profit organization) may be achieved by better cooperation in the chain. It imposes demands on one's own organization (one's own system), forcing work to be done on internal processes and professionalism, and ultimately leading to improved services/products.

[2]These phases were developed in the eighties under the leadership of Hardjono into the so-called *Berenschot Generation Model* and among other things described in Hardjono and Hess (1993) *De Nederlandse kwaliteitsprijs en onderscheiding, Kluwer, Deventer.*

Also, Fisscher (1994) alludes to the companies' responsibility toward society. "Companies are regionally, nationally and globally important actors who, aside from political decision makers, exercise much influence on a variety of areas. Whether in terms of employment, quality of work, price levels ranging from raw materials to final product, environmental footprint, social relations, direct or indirect, companies have a significant impact on prosperity and well-being, now and in the future". Fisscher (2015) mentions the importance of what he refers to as *fair quality.* "Fair quality shows itself from the inside out and is value-oriented, in an organization willing and able to be accountable". And: "Integrity may be described as acting based on a sound judgment and on searching for a unity or wholeness of thinking and acting. Essential for *integrity* is the combination of soundness and wholeness/cohesion. Values and behavior must match one another" (Fisscher, 2017).

## 2.8 The Importance of the Context

Besides inside the organization, we must also look at the outside more. The idea that the context is crucial within quality management is not new. For instance, it was high on the agenda of the contingency theory (see, e.g. Woodward 1958). The *contingency theory* states that Taylor's scientific management is flawed due to the fact that it does not consider the surrounding factors affecting the organization and its leadership. Woodward (1958) pointed out that there is not one single best way of organizing. Similarly, there is not one single best way of providing leadership (cf. also Hersey et al, 2001 and their situational leadership).

Morgan (2006) states among other things that organizations are open systems and must adapt to external circumstances. Different environments require different organizations. Differing circumstances require that organizations opt for different approaches to quality. *"No one has the monopoly of the truth",* as van Kemenade (2015a, b) wrote about the arrogance of some Western quality experts.

The context provides meaning. Chapter 9 in its entirety deals with the importance of context.

## 2.9 Affirmation, Negation, Compensation, and Solidarity

But how should we respond to the context? If we take a side-step, in the sociology of literature the concept of *function* of a literary work is applied in the sense of *intended or desired effect.*

In *Die Hauptrichtungen der Literatursoziologie. Ein Beitrag zur literatursoziologischen Theorie* (Bouvier, Bonn, 1964), the author Fügen distinguishes three types of literature, based on their effect:

1. The society-conforming type (aiming at maintaining the existing order, that is stable and closed;
2. The society-opposing type (aiming at changing the existing order);
3. The society-averted type (adopting a hostile or indifferent attitude toward the existing order).

Vanheste (1981) in his *Literatuursociologie: Theorie en methode* expands this typology. In a diagram, he distinguishes the following three options (of a literary work to respond to reality): *Affirmation* (the literary work confirms the historical reality or facets thereof and strives to maintain or preserve reality); *Negation* (the literary work denies or resists the existing historical reality or facets thereof and strives to change existing reality); and *Compensation* (the literary work rejects reality aspects, adopts the point of view that they are unsatisfactory and unacceptable, and experiences them nonetheless as unchangeable).

Within these three main categories Vanheste distinguishes the following sub-categories:

Affirmation (confirmation of reality)

- By *distortion* (reality is presented in a distorted manner, violated, presented in an untruthful manner, etc.);
- By *concealment* (less pleasant facets of reality are marginalized or disguised; positive facets are strongly accentuated and idealized);
- By *registration* (reality facets are confirmed based on their actual importance).

Negation (resisting of reality)

- By confirmation of an *alternative:* a desirable and realizable future reality is conjured up (this may be a utopia, a reality from the future or from the past);
- By *unmasking*: the description of reality shows how great the shortcomings and abuses are;
- By *criticism*: shortcomings are not only registered, they are also denounced;
- By *resistance:* aspects of reality are not merely subjected to criticism, what follows is a call to actual resistance.

Compensation (rejection of reality without an alternative)

- By *escape*: reality aspects are disconnected from total reality and presented as a dream world;
- By *isolation*: a reality aspect is disconnected from totality and expanded into a (realistic) island where life is good;
- By *revolt*: reality is experienced as reprehensible, immutable, and inescapable. Fleeing is not possible, not in a dream nor in a form of isolation. What remains is the revolt: a powerless outburst of anger, the fierce rage against something that one cannot get away from.

van Kemenade (1976) added a type of a different order to this, a stirring type: solidifying literature, which may show solidarity both with existing reality and with the fighters of existing reality.

Although the typology has been designed within literary sociology, it offers opportunities to refer to other social movements as a certain response to reality, as a reaction to the context. Or they offer the means to analyse the answer given by a particular organization in response to changing reality.

*Typology offers possibilities to describe how an organization responded to major social changes.*

Organizations may deny, confirm, fight, or reject reality without an alternative. Quality is then an emergence view of the desired functioning of an organization in response to society. You may also think of corporate social responsibility,[3] and of fighting reality by confirming an alternative. Hardjono and Klein (2004) provide an example of the way in which various social values lead to different *qualities* in their European Corporate Sustainability Framework (ECSF). "The European Corporate Sustainability Framework (Hardjono) is based on the concept of phase-wise development by distinguishing different levels of existence and related interpretations of Corporate Responsibility and Corporate Sustainability. This distinction is seen as important, because of the character of the subjects, which are ambiguous. Corporate Sustainability and Corporate Responsibility always mean something, but nearly always something different to different people. By addressing these subjects consistently and in order to enable discussion a distinction between four different kinds of interpretations of Corporate Sustainability and Corporate Responsibility has been made (Werre and Van Marrewijk, 2003), based on the levels of existence as distinguished by Prof. Graves".

## 2.10 Values and More Definitions

Via the reaction of organizations to the context we now come to values. For indeed, whether you want to deny, confirm, fight, or reject reality without an alternative is strongly determined by values. We defined quality among other things as *the values* attributed by stakeholders to an entity. We define value as: ***"the set of established principles, governing virtuous behavior".***[4]

We are not only talking about "value for money" (Garvin, 1984, Feigenbaum, 1951, Harvey & Green, 1993, Sherkenbach, 1993, Drucker, 2000). Reeves and Bednar (1994) mention—aside from *quality is excellence*—that *quality is value* is their second definition of quality. And although they also provide two other definitions, it always concerns the same notion: The value which an interested party attributes to what the organization provides. Bollaert (2014) provides as a definition of quality: *quality is the added value between input and output.* Even Conti (2006) expresses it: *We say quality, we mean value, (o.c., p. 16).* In order to know what a person considers quality, it is important to know the values held by this person. The

[3]More about this in paragraph 1.4., as it is dealing with values.

[4]Read more: http://www.businessdictionary.com/definition/ethical-values.html.

same thing applies to an organization or a system. Where behavior and body language are the manners in which a person acquires values, it is the organiza2007tional culture with its agreements, rituals, and procedures that teaches an employee the organizational values. This thought also fits the conclusion of Weggeman () that professionals in organizations cannot be controlled by rules and procedures, but that it is important to create *shared values.*

Feigenbaum (1951) observed: *"Quality does not have the popular meaning of best in any absolute sense. It means best for certain customer conditions. These conditions are (a) the actual use and (b) the selling price of the product. Product quality can not be thought of apart from product cost."* Quality here amounts to the user's satisfaction per dollar.

We quoted Peter F. Drucker before as saying: *"It is what the customer gets out of it and for which he is willing to pay"*. Earlier, Deming mentioned the importance of the cost: *"Quality is continuously meeting the customers' needs and expectations, at a price they are willing to pay"* (in Sherkenbach 1993). In the terminology of Garvin (1984), it concerns his value-for-money approach. *"Quality is the degree of excellence at an acceptable price and the control of variability at an acceptable cost"*. Moreover, in the eyes of Garvin, price not only concerns money but also the efforts a customer makes in order to obtain a service.

Pirsig goes much further than Garvin. It concerns values in general, not only financial values. Having been inspired by Pirsig (1991) argue for introduction of a broader value concept in the discussion of the definition of quality. They are not the first in this respect. Reeves and Bednar (1994) mention—aside from "quality is excellence"—"quality is value" as their second definition of quality. And although they also provide two other definitions, it always concerns the same notion: The value which an interested party attributes to what the organization provides. Even Conti (2006) expresses it: *"We say quality, we mean value"*. In order to know what a person considers quality, it is important to know the values held by this person. The same thing applies to an organization or a system. Where behavior and body language are the manners in which a person acquires values, it is the organizational culture with its agreements, rituals, and procedures that teaches an employee the organizational values. This thought also fits the conclusion of Weggeman (2007) that professionals in organizations cannot be controlled by rules and procedures, but that it is important to create *shared values.* Fisscher (2015) stated above that (fair) quality is *value-based.*

The EFQM model defines values as: *"the understandings and expectations that describe how organization's people behave and upon which all business relationships are based (e.g. trust, support and truth)"*.

Robbins (1991) emphasizes the importance of values as controlling our behavior. *"We need to realize that the direction of our lives is controlled by the magnetic pull of our values. They are the force in front of us, consistently leading us to make decisions that create the direction and ultimate destination of our lives. This is true, not only for us as individuals, but also for the companies, organizations, and the nation of which we are a part"*.

Kok (1999) enunciates it as follows: *"By a value, someone is moved within the limitations of his nature, in a double manner: he becomes emotional about something and decides to act"*.

This thought relates to the neurological levels of Dilts (1996). Inspired by the work of Bateson (1972), he distinguishes six levels ranging from abstract to concrete, determining the behavior of humans and organizations (Table 2.3):

According to Dilts (1996) values are *"an important drive for convictions (and behavior)"*. Therefore, they are also important for the vision of the interested parties regarding an entity's quality. One could argue quality also has a spiritual aspect (De Vries, 1999; Zohar, 2000; Hardjono & Klamer, 2005).

De Vries (1999) provides the definition: Quality is fitness for purpose, meeting what's intended. "*Quality care as fitness for purpose may be implemented per aspect by finding the rules and normative criteria for the company*". He refers to Dooyeweerd (1953).

De Vries mentions fifteen aspects (Table 2.4):

**Table 2.3** Dilts (1996) neurological levels

| | |
|---|---|
| 1. Why I am | My mission |
| 2. Who I am | My identity |
| 3. My *belief system* | My values, motivation, convictions |
| 4. My skills | My capacities |
| 5. What I do | My behavior: my actions and reactions |
| 6. My context | The external context with limitations and opportunities |

**Table 2.4** Fifteen aspects of fitness for purpose of De Vries (1999)

| | |
|---|---|
| 1. | Arithmetic aspect (number aspect) |
| 2. | Spatial aspect |
| 3. | Kinematic aspect (movement aspect) |
| 4. | Physical aspect |
| 5. | Biotic aspect (environment) |
| 6. | Psychic aspect |
| 7. | Analytical aspect |
| 8. | Cultural historical aspect |
| 9. | Language aspect (communication) |
| 10. | Social aspect or relationship aspect |
| 11. | Economic aspect |
| 12. | Aesthetic aspect |
| 13. | Legal aspect |
| 14. | Ethical aspect |
| 15. | Faith-related aspect |

## 2.11 Quality According to Deming

Deming describes quality as: *A way of managing the business organization* (1986). Deming gave *fourteen key principles for management to follow* to significantly improve the effectiveness of a company or an organization.

In Deming's fourteen points, see Table 2.5, we find some arguments for such a holistic view and constantly shifting judgment depending on the most recent insights. Insights that need to be soundly determined, with no eternally valid statements,

**Table 2.5** Deming W. E. (1986) Out of the crisis, MIT Press, Cambridge, MA, USA

| |
|---|
| 1. Create constancy of purpose toward improvement of product and service, with the aim to become competitive and to stay in business, and to provide jobs |
| 2. Adopt the new philosophy. We are in a new economic age. Western management must awaken to the challenge, must learn their responsibilities, and take on leadership for change |
| 3. Cease dependence on inspection to achieve quality. Eliminate the need for inspection on a mass basis by building quality into the product in the first place |
| 4. End the practice of awarding business on the basis of price tag. Instead, minimize total cost. Move toward a single supplier for any one item, on a long-term relationship of loyalty and trust |
| 5. Improve constantly and forever the system of production and service, to improve quality and productivity, and thus constantly decrease costs |
| 6. Institute training on the job |
| 7. Institute leadership. The aim of supervision should be to help people and machines and gadgets to do a better job. Supervision of management is in need of overhaul, as well as supervision of production workers |
| 8. Drive out fear, so that everyone may work effectively for the company |
| 9. Break down barriers between departments. People in research, design, sales, and production must work as a team, to foresee problems of production and in use that may be encountered with the product or service |
| 10. Eliminate slogans, exhortations, and targets for the workforce asking for zero defects and new levels of productivity. Such exhortations only create adversarial relationships, as the bulk of the causes of low quality and low productivity belong to the system and thus lie beyond the power of the workforce<br>Eliminate work standards (quotas) on the factory floor. Substitute leadership<br>Eliminate management by objective. Eliminate management by numbers, numerical goals. Substitute leadership |
| 11. Remove barriers that rob the hourly worker of his right to pride of workmanship. The responsibility of supervisors must be changed from sheer numbers to quality |
| 12. Remove barriers that rob people in management and in engineering of their right to pride of workmanship. This means, *inter alia*, the abolishment of the annual or merit rating and of management by objective |
| 13. Institute a vigorous program of education and self-improvement |
| 14. Put everybody in the company to work to accomplish the transformation. The transformation is everybody's job |

but a basis for improvement. Whether or not affected by Deming's thoughts, this emergence thinking constitutes Japanese thinking about quality.

## 2.12 The Stakeholders

We defined quality as the value that *interested parties or stakeholders* attribute to an identity. ISO 9000:2005 provides the following definition of quality: "The totality of properties and characteristics of a product or service of importance for meeting the determined and obvious needs". The needs are directly related to a stakeholder, for example, the user. Even Garvin (1984) distinguishes the user-oriented approach. Oakland (1993) speaks about "meeting customer requirements". Also within this approach fits Juran (1980), who defines quality as: "the fitness for use", as do Gryna, and Bingham (1974). They assert that quality does not happen by accident but that it results from planning. Feigenbaum (1999) states that: *"the key is transforming quality from the past emphasis on things gone wrong for the customer to emphasis upon the increase in things gone right for the customer, with the constant improvement in the sale and revenue growth"* (p. 376). In this context, John Guaspari gives a paraphrase to the transcendental approach of quality: *Quality? I know it when I see it.* Which he changes into: *Quality: You will know it when* ***they*** *see it.*

As Cuijvers (1985) rightly so observes, there are two aspects to this definition. Quality is measurable, and secondly, the norm applied for the assessment of the product, the service, or the person is determined by the user (and not by the producer). In the same tradition, several other authors mention the customer as the one determining quality.

> Quality in a product or service is not what the provider puts into it, it is what the customer gets out of it and for which he is willing to pay.

Also Drucker comments: "*Quality in a product or service is not what the provider puts into it. It is what the customer gets out of it and for which he is willing to pay*". The highest quality has the product or service that best meets the expectations of a particular customer (group of customers).

Deming observes that quality goes even further than that: "*Quality should be aimed at the needs of the customer, present and future*" (in Sherkenbach 1993). In another application of the thought that quality is more than *meeting the customer's needs*, the bar is even raised higher: it is about surprising the customer with more than he requests. Reeves and Bednar (1994) mention: *meeting and/or exceeding customer's expectations.* We find the same thought with others (Gronroos, 1982; Parasuraman, Zeithaml, & Berry, 1985).

> Quality is so extraordinary that it delights the customer (Kano, Nobuhiko, Fumio, & Shinichi, 1984).

It is about achieving the customer's loyalty. This definition of quality fits with the Reference Paradigm.

We prefer to use the broader term *stakeholders*, instead of *user*, or *customer*. Vanhoof and Van Petegem (2007) define quality of education as: *meeting the expectations of the stakeholders in a suitable manner*. A widely accepted, broad definition of *stakeholder* is the one coined by Freeman (1984, p. 46), who considers a stakeholder to be: *any group or individual who can affect or is affected by the achievement of the organization's objective*. And this results in many stakeholders (De Vries, Verheul, & Willemse, 2003). Quality is a political issue. Who are all interested parties? Whose needs do we want to meet? Whose expectations are intended? Who determines what is quality? Quality is then per definition a subjective notion, *"quality lies in the eyes of the beholder"*. We see that the various stakeholders make different choices. So quality is a relative concept, related to the user's judgment. The user determines the product and expresses his opinion about it, for example, in terms such as good, average, or poor quality.

The question whom we should listen to is important when companies discuss their quality. In the INK management model, we see that a balance must be struck between *customer appreciation*, *employee appreciation*, *community appreciation*, and *management appreciation*. "As an organization is able to do more, the *value* of the organization grows. The value will be different for each interested party (shareholders, management, employees, customers and other stakeholders) and will also be expressed in other dimensions (market capitalization, sales value, status value, value as living- and working environment, value as supplier, value as customer, news value etc.)" (Hardjono 1995).

*Each of us brings different priorities to discussions of quality*

We see that there are various stakeholders together (not only customers, personnel, management, and shareholders but also suppliers and partners) each with their own norms. Or in the words of Levine: *"Each of us brings different priorities to discussions of quality"* (Levine 1982). This is indeed the risk if quality is a subjective concept. Harvey and Green (1993) also reach this conclusion where the quality of higher education is involved: *"At best perhaps, we should define as clearly as possible the criteria that each stakeholder uses when judging quality and for these competing views to be taken into account when assessments of quality are undertaken"* (o.c., p. 28). Some writers feel that determining quality due to the many stakeholders is not possible.

## 2.13 Conclusions

In the paragraphs above, we introduced various definitions of quality (a.o. Juran, Crosby, Feigenbaum, Garvin, 1984; Harvey & Green, 1993; Reeves & Bednar, 1994). There is no single definition of quality that is generally applicable. The question is on what people base this truth. Is a Rolls Royce still the best car if the price is taken into consideration? Feigenbaum (1951) already mentioned that: *quality does not have the popular meaning of best in any absolute sense*. Even Crosby (1979) indicates how hard it is to define quality: ".... *In this regard, quality has much in common with sex. Everyone is for it (under certain conditions of course.). Everyone feels they understand it. (Even though they wouldn't want to explain it.) Everyone thinks execution is a matter of following natural inclinations. (After all, we do get along somehow.) And, of course, most people feel that all problems in these areas are caused by other people. (If only they would take time to do things right*)".

The various definitions apply to different circumstances (Reeves & Bednar, 1994; Sousa & Voss, 2002). Moreover, the "multi-dimensional character of quality" could not be emphasized enough (Sousa & Voss, 2002). Therefore, we reach the following conclusions until then:

1. *Quality is the capability, nature of something/* of an entity, but apparently also *the degree in which something is good.*
2. Depending on one's paradigm perspective, quality differs. Quality differs also according to the sector of society. In the services sector quality is not merely a thing, but quality is part of an event, the interaction.
3. Quality can not only be considered as a relative concept but also as a transcendent concept.
4. The definition of the quality concept depends on the paradigm's perspective.
5. Pirsig has been cited much too often to indicate that quality cannot be defined. In his second book, *Lila*, he certainly made some contributions, leading to a dynamic quality concept, fitting the Emergence Paradigm.
6. The quality concept also depends on the context.
7. Garvin provides eight characteristics of a product, while additionally mentioning the characteristics of a service.
8. Process, profession, the organization/system, the chain, but also society may all be entities of which one needs to determine and/or improve the capability.
9. Deming's fourteen points are a holistic view of quality. His view better fits in the Emergence Paradigm than other paradigms.
10. The manner in which quality management has been introduced by Deming in Japan has been misinterpreted until now.
11. The context is an important factor regarding the questions what, where, and when quality exists.
12. Affirmation, negation, compensation, and solidation are typologies offering possibilities to explain movements of organizations as a particular response to reality, in a given context.
13. The concept *quality* must be linked to the concept *values*, which *stakeholders* attribute to an entity.
14. Values control our behavior, which may be divided into six levels: my mission, identity, values, motivation, convictions, capabilities affecting my behavior. My behavior in turn is determined by my actions and my surroundings, the external context with its limitations and opportunities.
15. Quality may also be regarded as: *fitness for purpose* (De Vries), to which Dooijeweerd's fifteen aspects may be linked when we attempt to determine rules and normative criteria.

16. "Quality in a product or service is not what the provider puts into it. It is what the customer gets out of it and for which he is willing to pay".
17. Each of us brings different priorities to the discussion of quality.

## References


Bateson, G. (1972). *Steps to Ecology of Mind*. New York: Ballantine Books.

Bobbink, H. (2010). 's-Lands Wijs, 's-Lands Eer. *Sigma, 2,* 12–16.

Bollaert, L. (2014). *A manual for the internal quality assurance in higher education*. Brussels: Eurashe.

Crosby, P. B. (1979). *Quality is free*. New York: McGraw-Hill.

Cuyvers, G. (1985). *Integrale kwaliteitszorg*. Leuven: Acco, pagina.

De Vries, H. J. (1999). *Kwalitetszorg zonder onbehagen*. Amsterdam: Buijten & Schipperheijn.

De Vries, H. J., Verheul, H. H. M., & Willemse, H. (2003). Stakeholder identification in IT standardization processes. In J. L. King & K. Lyytinen (Eds.), *Proceedings of the Workshop on Standard Making: A Critical Research Frontier for Information Systems* (pp. 92–107). Seattle: International Conference on Information Systems.

Deming, W. E. (1986). *Out of the crisis*. Cambridge Mass: MIT Press.

Dilts, R. B. (1996). *Visionary Leadership Skills*. Capitola: Meta Publications.

Dooyeweerd, H. (1953–58). *A new critique of theoretical thought, Vol. I–IV*. Ontario: Paideia Press (1975 edition).

Drucker, P. (2000). *Management*. Amsterdam: Antwerpen, publishing company Business Contact.

Feigenbaum, A. V. (1951). *Quality control: Principles, practice, and administration*. New York: McGraw-Hill.

Feigenbaum, A. V. (1999). The new quality for the twenty-first century. *TQM Magazine, 11*(6), 376–383.

Fisscher, O. A. M. (1994). *Kwaliteitsmanagement en bedrijfsethisch handelen*. Enschede: Universiteit Twente.

Fisscher, O. A. M. (2015). Integere kwaliteit, *Sigma, 2,* 12.

Fisscher, O. A. M. (2017). Integriteit en Innovatie: dilemma's uit de praktijk. *Sigma, 1,* 24–27.

Freidson, E. (2001). *Professionalism*. The Third Logic: Polity Press, Cambridge, UK.

Garvin, D. A. (1984). What does 'product quality' really mean? *Sloan Management Review*, 25–43.

Garvin, D. A. (1987). Competing on the eight dimensions of quality. *Harvard Business Review*. Retrieved December 20, 2016.

Gilmore, H. L. (1974). Product conformance cost. *Quality Progress, 7*(5), 16–19.

Gronroos, C. (1982). *Strategic management and marketing in the service sector.* Helsingfors: Swedish School of Economics and Business Administration.

Hardjono, T. W. (1995). *Ritmiek en organisatiedynamiek*. The Hague, Kluwer: Vierfasenmodel.

Hardjono, T. W., & Bakker, R. J. M. (2002). *Management van processen*. Deventer: Kluwer. ISBN 9013034446.

Hardjono, T. W., & De Klein, P. (2004). Introduction on the European corporate sustainability framework (ECSF). *Journal of Business Ethics, 55*(2), 99–113.

Hardjono, T. W., & Hess, F. W. (1993). *De Nederlandse Kwaliteitsprijs en Onderscheiding*. Deventer: Kluwer.

Hardjono, T., & Klamer, H. (2005). *Breng spirit in je werk! Hoe doe ik dat?*. Zoetermeer: Meinema.

Harteloh, P. P. M. (2000). *Het begrip kwaliteit in de gezondheidszorg*. Rotterdam: Erasmus Universiteit.

Harvey, L., & Green, D. (1993). Defining quality. *Assessment & Evaluation in Higher Education, 18*(1), 9–34.

Hersey, P., Blanchard, K. H., & Johnson, D. (2001). *Management of organizational behaviour: Leading human resources*. Prentice Hall.
Juran, J. M., Gryna, F. M., Jr., & Bingham, R. S. (Eds.). (1974). *Quality control handbook* (3rd ed.). New York: McGraw-Hill.
Kano, N., Nobuhiko S., Fumio T., & Shinichi T. (1984). Attractive quality and must-be quality. *Quality: The Journal of the Japanese Society for Quality Control, 14*, 39–48.
Kelchtermans, G. (2003). School in balans. *Proceedings COV-conferentie 2002*. Schoolwijzer: Antwerpen.
Kok, J. F. W. (1999). *Normen, waarden en waarden-loze normering*. Baarn: Nelissen.
Levine, A. E. (1982). Quality in baccalaureate programs: What to look for when David Riesman can't visit. *Educational Record*, 63, 13–18.
Levitt, T. (1972). Production-line approach to service. *Harvard Business Review, 50*(5), 41–52.
Marrewijk, M., & Werre Van, M. (2003). Multiple levels of corporate sustainability (CS). *Journal of Business Ethics*.
Oakland, J. S. (1993). *Total quality management, the route to improving performance*. Oxford: Butterworth-Heinemann Ltd.
Pirsig, R. M. (1991). *Lila: An inquery into morals*. London: Transworld Publishers Ltd. ISBN 0-593-02507-5.
Reeves, C. A., & Bednar, D. A. (1994). Defining quality: Alternatives and implications. *Academy of Management Review, 19*(3), 419–445.
Sherkenbach, W. W. (1993). *Met Deming op weg naar kwaliteit en produktiviteit*. Deventer/Antwerpen: Kluwer.
Sower, V. E., Motwani, J., & Savoie, M. J. (1995). *Classic Readings in Operations Management Ft. Worth, Texas*. The Dryden Press, pp. 191–214.
Stevens, F. Bering R., & Stevens, K. (1999). *Leren excelleren* 12 Jaar General electric tegen de achtergrond van het INK-model.
Taguchi, G. (1992). *Taguchi on robust technology development: Bringing quality engineering upstream*. ASME Press. ISBN 978-0791800287.
van Kemenade, E. (1976). *De Functie van Politiek Toneel*, masterscriptie Literatuursociologie, Katholieke Universiteit Nijmegen.
van Kemenade, E. (2015a). De arrogantie van de Westerse kwaliteitszorg. *Sigma* 3, june 2015, pp. 6–10. Retrieved 4th of March from https://www.researchgate.net/publication/315462396_De_Arrogantie_van_de_Westerse_Kwaliteitszorg.
van Kemenade, E. (2015b). Contextual divide. An organization's context should determine leadership approach. *Quality Progress*. 2015 November. Retrieved 4th of March from http://www.linkedin.com/in/everardvankemenade.
Vanhoof, J., & Van Petegem, P. (2007). *Interne en externe evaluatie in een context van verantwoording en schoolontwikkeling. Hoe verenigbaar zijn beide? ORD 2007*. Groningen: RUG.
Vanheste, G. (1981). *Literatuursociologie: Theorie en methode*. Van Gorcum.
van Zoest, C. (2001). *Kwaliteitszorg voor non-profitorganisaties*. Baarn: H. Nelissen.
Vinkenburg, H. H. M. (1995). *Stimuleren tot perfectie; kritieke factoren bij het verbeteren van dienstverlening*. Deventer: Kluwer bedrijfswetenschappen.
Vinkenburg, H. (2007). Open brief van Huub Vinkenburg op de vorige Synaps. In *Synaps* (Vol. 22).
Weggeman, M. (2007). *Leidinggeven aan professionals? Niet doen!* Schiedam: Scriptum.
Woodward, J. (1958). *Management and technology*. London: Her Majesty's Stationary Office.
Zohar, D. (2000). *Spiritual intelligence: The ultimate intelligence*. Bloomsbury paperbacks.

# Chapter 3
# Classification of Thinking About Quality

> Thinking must deny itself.
>
> Krishnamurti: The world that is you.

In our introduction, we previously mentioned the clearly discernible schism in the thinking about quality in the eighties of the previous century. Possibly not all of the parties involved were aware of that when they pleaded for more attention being given to quality. New ideas gained traction on how quality ought to be considered and particularly the *quality management* concept emerged. Prior to this period the discussion is more about *quality control*. Such ideas are specifically described in the form of paradigms, schools, phases, and elements. Thinking about quality in the form of paradigms, on which our book is based, has to pay tribute to these views and constitutes a sufficient reason to include these views in our publication. They also constitute a source of our thoughts and contribute to the basis of the choices we made. We have abandoned some views, while other views have been partially included or excluded. By means of these views we will investigate the question: "What might be the characteristics of a new paradigm in quality management?"

## 3.1 Vinkenburg's Quality Schools

> It may be quite plausible that "excellence" in general and the quality of provided service in particular are based on three pillars.

One of the reasons for our book is the discussion Huub Vinkenburg initiated in the introduction of his three schools. Vinkenburg (2006) states that managers and advisors should be looking up to philosophers; they have an all-encompassing model (Kunneman, 1985). He shows how reality may be subdivided into three domains:



T. Hardjono and E. van Kemenade, *The Emergence Paradigm in Quality Management*, https://doi.org/10.1007/978-3-030-58096-4_3

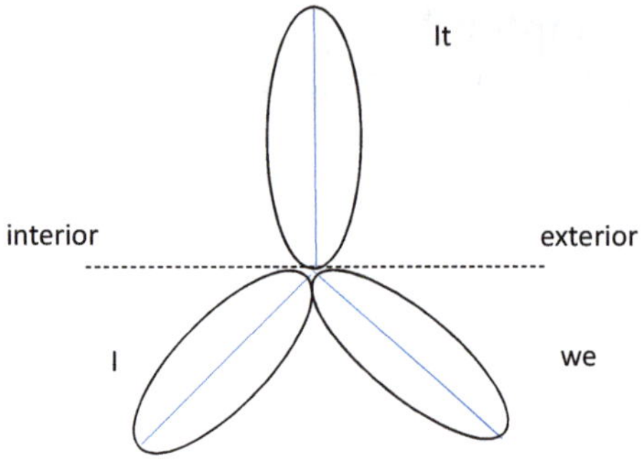


**Fig. 3.1** The three domains according to Wilber (2000)

nature, humanly inward, and culture—we henceforth referred to this as the It, I, and We domains, respectively. *It* is objective, *I* is subjective, and *We* is inter-subjective.

Based on his model, it is quite plausible that *excellence* in general and the quality of provided service in particular are based on three pillars.

- The service (It)
- The service provider's attitude (I)
- The interaction between service provider, customer, and manager (We).

Compare Pirsig for the latter aspect, who states: "Quality is not a thing, it is an event". At a glance, according to Vinkenburg, you can now discern what is missing in the quality science philosophy. *"It is not aimed at the inward, the unmeasurable, the subjective, the unmakable, the uncontrollable. It does not deny the existence and influence of the human factor but is not able to do anything with it. It has empty hands with (open) problems in the I and We area, such as those occurring in case of a 'wrong attitude' and 'unfruitful interaction'.*

The same three domains conform with Wilber (2000)[1] (see Fig. 3.1).

Based on the three domains, Vinkenburg (2007) distinguishes three schools, each with their own concept of *quality.* In his publication *Dienstverlening; paradigma's, deugden en dilemma's* (Services; Paradigms, Virtues and Dilemmas) of 2006 he mentions virtue-based ethics. He regards (subjective) reflection as a means to achieve virtuous actions within organizations. In 2007, he still utilizes the division into the statistical, the business administration, and the reflexive school. "Objects from the *It* domain can be managed from Control, whereas objects from the *I* and *We* domains may be influenced from Involvement". In later publications (2009, 2010, 2012), he utilizes three other names in conjunction with a different explanation. Now he distinguishes between the Empirical School, the Normative School, as well as the Reflective School (see Table 3.1).

[1] He bases himself on a simplification which Wilber himself applies in his original model with four quadrants:

"The four quadrants can be simplified to the "Big Three" (I, we and it)" (Wilber, 2000; p. 52). However, in a later stadium Vinkenburg emphatically distances himself from Wilber and admits regretting having adopted his ideas.

**Table 3.1** The three schools according to Vinkenburg (2006, 2007)

| Wilber, Vinkenburg (2006, 2007) | | | | | | |
|---|---|---|---|---|---|---|
| Aspect | Paradigm | School's nature | Quality concept characterization | Entity | Benchmark | Requirements |
| *Outward/Individual (It)* | *Control Paradigm* | *1. Statistical School* | Quality is the inescapable result of a controlled process. (Vorstman) | The product, primary process | Objective | True, effective |
| *Inward collectively (We)* | *Involvement paradigm* | *2. Business Administration School* | Quality is the degree to which a totality of properties and characteristics meets the requirements. (ISO 9000) | The organization as a system | Intersubjective | True, real |
| *Inward/ Individual (I)* | | *3. Reflexive School* | Quality is an event that hits the spot, and that contributes to the quality of life. | The group and human as individual | Subjective | Correct, valid |

Vinkenburg has always avoided a fourth school. In his argumentation he asks the questions: What would be the questions and problems of such a fourth school? What are the symptoms that the school tries to manage, its diagnosis, therapy, and treatments? Vinkenburg states that more paradigms are not necessary. Yet, Hardjono (2010), Kemenade and Hardjono (2011), Kemenade (2011, 2012, 2014a, b), Schijndel and Kemenade (2011) wondered whether there should be a fourth school and what it should include. Previously touched upon in the introduction, Peter Noordhoek in his thesis *Trusting Associations* (2019) demonstrates that the three known approaches do show results. However, they do not function to full satisfaction. The approaches that he describes exhibit some major similarities with the abovementioned three schools. In his case descriptions of associations, he demonstrates that in the course of time, by pressure from the context and the willingness to reach a consensus about what must be understood as quality. Yet another approach appears to evolve. We keep on searching.

## 3.2 The Revolutions in Quality Thinking by Bertrand Jouslin de Noray

From 2001, a discussion started within the European Organization for Quality (EOQ) about the different ways to regard quality management. Bertrand Jouslin de Noray, then Secretary General of EOQ, led the discussion. The breeding places for the discussion were the Summer Camps. The first EOQ Summer Camp took place in Versailles, in July 2001, with 60 Change Leaders from 13 countries present. Subsequent Summer Camps were organized in Sweden, Estonia, Lebanon, and The Netherlands. Dutch delegates played an active role, taking the initiative to organize Summer Camps especially for health care. Since 2007, a Dutch variant of the Summer Camp traditionally takes place in January, therefore referred to as *Winter Camp*. The organizer is the Dutch Academy for Quality represented by Arend Oosterhoorn and Teun Hardjono. The subsequent series of topics portrays a beautiful example of emergence thinking about quality: "Quality of Humans, Organization and Society" (2007); "Worlds of Difference. Performance, Paradigms and Models" (2008); "Quality of the Unmeasurable and the Unmeasurability of Quality" (2009); "It Must Be Correct!" (2010); "A New Grip!" (2011); "The Limit of the Unknown" (2012); "Passion" (2013); "Changing Society Asks for Changing Quality/Qualities" (2014); "Non-violent Organization, the Desire for the Endless Sea" (2015); "Quality (Management) in Space and Time" (2016); "Quality Management in Organizations as Shared Spaces" (2017).

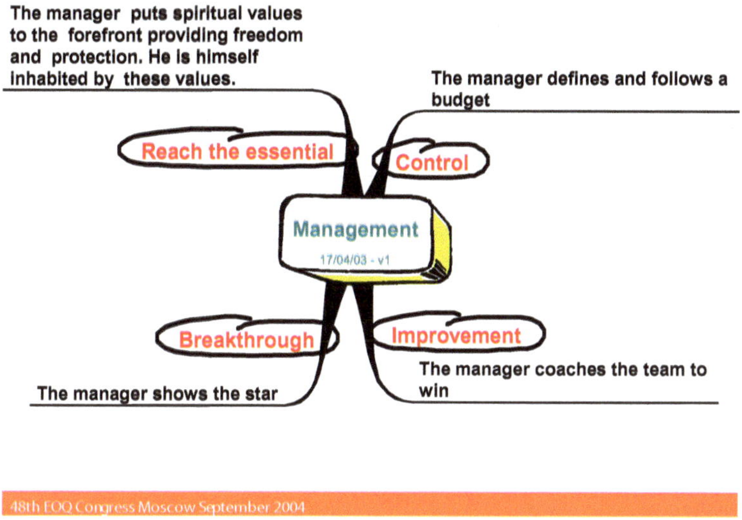


**Fig. 3.2** Revolutions according to Jouslin de Noray (2004)

The summer and winter camps serve as the backdrops to numerous congress presentations. In May and September 2004, De Noray gave a presentation, respectively, in Copenhagen and Moscow on the theme: "Theory and techniques on breakthrough change".[2] According to De Noray, the quality movement, as we know it, brought forth substantial progress to organizations, following World War II. Organizational changes were introduced over time. De Noray refers to these as *revolutions*, since the transition from the one to the other may not be as smooth. According to De Noray's idea, a paradigm shift is required and specific manners must be implemented to fluctuate from one quality vision to another. Yet he does not explain what is meant by a paradigm. This is something we will do later in our book.

He describes the following four revolutions in quality and change management (see Fig. 3.2).

1. Control
2. Continuous improvement
3. Breakthrough
4. Reaching the essential.

Jouslin de Noray (2004) calls (process) control the first revolution in quality management. An example of this movement is how, from the end of the nineteenth century onwards, the—then new—phenomenon of mass production addressed quality. The production process was divided into constituting processes which made the final inspection even more important. Workers were no longer responsible for the

[2]Jouslin de Noray (2004): "Theory and Techniques on Breakthrough Change", presentation on the 48th Congress of the EOQ, Moscow.

**Table 3.2** Differences between the four revolutions according to Jouslin de Noray

| Control | Continuous improvement | Reaching the essential | Breakthrough |
|---|---|---|---|
| • Logic<br>• Contract<br>• Guarantee<br>• Be conform<br>• Satisfaction of work done<br>• Protection<br>• Threat<br>• No initiative | • Proudness of winning a challenge<br>• Success<br>• Teamwork<br>• Do be sure and check the success | • Trust in unknown<br>• Lost control<br>• Unconceivable targets<br>• Peacefulness in community with other<br>• Do because you trust and feel in peace | • Listening<br>• Discovery<br>• Dream<br>• Openness<br>• Do the way you feel and listen to what happens<br>• Enthusiasm of reaching a dream |

totality of the product, but only for their partial task. They received detailed instructions on how to do their job. It is the era of scientific management (Taylorism). De Noray calls integral quality care the second revolution in quality management. De Noray explains the "revolutions" in more detail, as illustrated in Table 3.2.

De Noray found the various opinions so fundamentally different, that he refers to the "revolutions" as a true paradigm shift. Therefore, he refers to each form or phase as a *paradigm.* We want to emphasize that *reaching the essential* revolution shows major similarities to what we refer to as the Reflective Paradigm, control with the Empirical Paradigm, Continuous Improvement with the Reference Paradigm.

There is room for a fourth paradigm that he mentions *breakthrough.*

## 3.3 Shoji Shiba and Breakthrough Management

In all three cases, a company needs to make money to survive.

In April 2005, during the pre-conference of the 49th EOQ Congress on "Accreditation of Higher Education", Shiba expounded on his view of the first three paradigms: control, incremental improvement and breakthrough. In "Breakthrough management" Shiba and Walden (2006) deal with these three paradigms in greater depth. Jouslin de Noray stated that: "*In all three cases, a company needs to make money to survive*". In the control paradigm, making money requires process control and standardization. In the incremental improvement paradigm, making money requires customer satisfaction as well as a quality culture and a quality system. In the breakthrough paradigm, making money requires bringing a new value into the market and, therefore, it requires innovation. In control management, the standards do not change, but the processes do. In incremental improvement the standards change, but the business remains more or less the same. In breakthrough management, the values of the organizations do not change, but the business does. Control, in the management view, correlates with what McGregor calls Theory X: people want to be led; incremental

**Table 3.3** Shiba's breakthrough explanation

| | Control | Incremental improvement | Breakthrough |
|---|---|---|---|
| *Business logic* | Control process | Customer satisfaction | New market value |
| *Changing* | Process | Standards | Business |
| *Unchanging* | Standards | Business | Values |
| *Hypothesis of human nature* | Theory X | Theory Y | Theory X |
| *Key player* | Middle manager | Shop Floor workers | Top managers |
| *Management focus* | Discipline | Backward | Forward |
| *Data* | Numeric | Numeric and language | Language and image |
| *Tools* | Statistical analysis | Kaizen methods | Breakthrough methods |

improvement fits Theory Y, which says that people seek autonomy and innovation. In breakthrough, Theory Z is required and we need to go beyond rational thinking. In control, a key role must be played by middle managers. In incremental improvement the key role must be played by the workers on the shop floor. In breakthrough the key role must be played by the top managers. The management focus is, respectively, on discipline, the past, the future. Data utilized are numerical, numerical and linguistic, linguistic and image.

Instruments of control are statistical analyses, ranging from incremental improvement Kaizen methods and methods as visual thinking, point of view approach and the five-step-discovery process (Shiba & Walden, 2006, p. 36). Shiba's contribution was aimed among other things at explaining further details of the breakthrough concept (Table 3.3)

## 3.4 The Value Systems of Beck and Cowan

Of a different order, but not to be ignored in the classification of thinking about quality is the division into value systems. We already mentioned that we defined quality as the value which stakeholders attribute to an entity. The value systems in Spiral Dynamics of Beck and Cowan (1996) are excellent for characterizing quality science thinking. These are marvelously aligned with the four *revolutions* of Jouslin de Noray and Shiba (control, continuous improvement, breakthrough, and reaching the essence) described in the above paragraphs. Originally, Beck and Cowan describe many more paradigms, but we want to restrict ourselves here to a descriptive diagram of five value systems most important to us. These are:

1. Truth Force, with an emphasis on authority, morality, rules, tradition
2. Strive Drive, with an emphasis on success, growth, consumerism, opportunities
3. Human Bond, with an emphasis on community, authenticity, sharing, caring
4. Flex Flow, with an emphasis on systems, self-organization, networking
5. Whole View, with the emphasis on holism, cosmos, spirituality

Truth Force fits the Empiric Paradigm, Strive Drive fits the Reference Paradigm; Human Bond fits the Reflective Paradigm. FlexFlow and Whole View provide an opening to a new paradigm.

Laloux (2014) applies Spiral Dynamics to organizations. He provides a somewhat different division into paradigms compared to Beck and Cowan, yet also bases his views on the same source Graves. Laloux pleads for wholeness, listening to the evolving goal and self-determination in organizations. Among other things, he mentions the development which neighborhood care experienced in the Netherlands. Leadership does not determine the direction but listens to the direction in which an organization wants to move. It is not *command and control*, but *sense and respond (compare Cynefin in the next chapter).*

## 3.5 Conclusions

In the above-mentioned paragraphs we introduced various ways of thinking about quality (o.a. Jouslin de Noray, 2004; Shiba and Walden, 2006; Vinkenburg, 2006, 2007; Beck & Cowan, 1996; Laloux, 2014). The revolutions of Jouslin de Noray and Shiba, as discussed in the EOQ, definitely provided an impulse for our thinking. We were partially in agreement with the Vinkenburg's schools. Despite all objections one might have against the Beck and Cowan's value orientations, we were inspired by their ideas.

Both Jouslin de Noray as well as Shiba observe how the paradigms dominate at different points in time. Jouslin de Noray sees the beginning of the control paradigm with mass production in the forties of the previous century. For Japan, Shiba puts this a decade later. For Jouslin de Noray, continuous improvement starts from the seventies. Here, Shiba sees this take off in Japan a decade earlier. Breakthrough starts in the globalization period from 2000, respectively in the nineties. Jouslin de Noray sees reaching the essential from 2010.

Among other things, there is scientific criticism on the Spiral Dynamics concept at this point. Critics, such as Bauwens (2005), feel that the transitions in Spiral Dynamics are too often presented as universal. There are many differences in the ways in which human cultures have presented themselves over time. The claim that people have systematically changed on the basis of psychosocial dimensions, such as self-concept, as suggested in Spiral Dynamics over time, is not supported by anthropology, social sciences nor evolutionary biology. Over many years, Cowan has not been in touch with his co-author. Cowan states that Beck c.s. put too much emphasis on colors and on the difference between the first and the second level (transition from green to yellow). For this, Wilber refers to the database of Graves. Cowan points out that for Graves there were still hypotheses in need of further investigation. He designed the colors in just fifteen minutes to illustrate a PowerPoint presentation.

In naming the paradigms for looking at quality, we do not use the characterization of eras in value orientations. We do not consider a potential difference between first and second tier meaningful. To us, value orientations help to indicate the various ways to perform quality management and how it is determined by the various contexts.

Yet we do not see a fourth paradigm looming. The classifications only evoke more questions. How do the revolutions of Jouslin de Noray relate to the Vinkenburg schools? How do the two of these relate to the value orientations of Beck and Cowan?

We come to the following conclusions:

1. De Noray, Shiba, and Beck and Cowan offer their own views tailored to the complex, even rapidly changing current reality.
2. There are schools, value orientations, and paradigms that seem to be related one to another, but exactly how is unclear.
3. In the classifications we researched, we found none comprehensive and containing all possible means of viewing quality.
4. De Noray, Shiba, and Beck and Cowan provide openings to a fourth paradigm.
5. There is every reason to seek answers to the uncertainties and incompleteness evident in viewing quality. We will deal with this in the following chapters, as our search continues.

## References

Bauwens, M. (2005). *A critique of wilber and Beck's SD-integral*, P/I: Pluralities/Integration, no. 61: March 23.

Beck, D., & Cowan, C. (1996). *Spiral dynamics*. Malden: Blackwell Publishers.

Jouslin de Noray, B. (2004). Theory and techniques on breakthrough change. In *EOQ Congress 48th*, Moscow, EOQ.

Kunneman, H. (1985). *Habermas' theorie van het communicatieve handelen*. Meppel: Boom.

Laloux, F. (2014). *Reinventing organisations*. Brussels: Nelson Parker.

Noordhoek, D. P. (2019). *Trusting associations: A surgent approach to quality improvement in associations*, Ph.D. study, Tilburg University.

Shoji, S., & Walden, D. (2006). Breakthrough management: Principles, skills, and patterns or transformational leadership, confederation of Indian industry. ISBN-10: 8190356437, ISBN-13: 978-8190356435.

van Kemenade, E. (2011). Briefwisseling over scholen in kwaliteit. *Sigma* 5, oktober 2011. Retrieved 4th of March from https://www.vankemenade-act.nl/wp-content/uploads/2017/08/SIGMA05_Briefwisseling.pdf.

van Kemenade E., & Hardjono, T. W. (2011). Van vier paradigma's naar vier scholen en een Wintercamp. *Synaps*, *32*, pp. 28–34. Retrieved 4th of March from https://www.vankemenade-act.nl/wp-content/uploads/2017/08/Vier_paradigmas-vier-scholen3.pdf.

van Kemenade, E. A. (2012). Soft skills for TQM in higher education standards. *ASQ Higher Education Brief,5*(2). Retrieved 4th of March from http://www.linkedin.com/in/everardvankemenade.

van Kemenade, E. (2014a). Theory C: The near future of quality management. *The TQM Journal, 26*(6), 650–657. Retrieved 4th of March from https://www.researchgate.net/publication/266967522_Theory_C_the_near_future_of_Quality_Management.

van Kemenade, E. (2014b). The myth of the PDCA cycle in times of emergent change. In: *Conference Proceedings EOQ 2014*. Retrieved 4th of March from http://www.linkedin.com/in/everardvankemenade.

Van Schijndel, B., & Van Kemenade, E. (2011). *Hoe goed is het nieuwe accreditatiestelsel*? *TH@MA* 4, pp. 4–8.

Vinkenburg, H. (2006). Dienstverlening; paradigma's, deugden en dilemma's. In *Kwaliteit in Praktijk B1–5*.

Vinkenburg, H. (2007). Open brief van Huub Vinkenburg op de vorige Synaps. In *Synaps* (p. 22).

Vinkenburg, H. (2009). Stromingen, scholen en zienswijzen in de kwaliteitskunde. In *Synaps* (p. 29).

Vinkenburg, H. (2010). Naar een derde school in de kwaliteitskunde? *Synaps, 31*, 3–5.

Vinkenburg, H. (2012). Kwaliteit in balans. *Sigma, 1*, 32–35.

Wilber, K. (2000). *A theory of everything: An integral vision on politics, science and spirituality*. Shambala publications.

# Chapter 4
# On Value Orientations, Schools, and Paradigms

In the previous chapter we presented various classifications, comparing common opinions side by side and relating them one to another without establishing a linear relationship. We are searching for answers and for a new movement. The question in this chapter is whether we can combine the one and the other and whether it concerns paradigms, movements, or value systems.[1] Eventually, we distance ourselves from the concept of school while opting for paradigm.

We deliberately decide to discuss paradigms in thinking about quality

## 4.1 Value Orientation According to van Kemenade

In order to know what a person considers quality, it is important to know the values held by this person.

Based on the ideas of Jouslin de Noray, Shiba, Beck and Cowan, and Vinkenburg (2006), Van Kemenade has added a new classification into value orientations. Particularly, he integrated Vinkenburg's control and involvement paradigm with the phases of Beck and Cowan, Shiba, and Jouslin de Noray.

Vinkenburg puts the control paradigm in opposition to the involvement paradigm that is particularly crucial in providing services, but not only there. "*Quality science-based thinking has a rational character. It grew in industry and was colored by the control paradigm. The field of expertise has a blind spot: it ignores the human factor, whereas this is essential in providing services. Additions to the quality science philosophy (may) be drawn from the involvement paradigm and virtue-based ethics.*" See Table 4.1. This not only concerns the relationship between manager and employee but also and specifically the relationship between employee and customer.

---

[1] We discuss the concept *value system* in the chapter on context.



T. Hardjono and E. van Kemenade, *The Emergence Paradigm in Quality Management*, https://doi.org/10.1007/978-3-030-58096-4_4

**Table 4.1** Characteristics of the two foundational paradigms according to Vinkenburg

| Control Paradigm | Commitment paradigm |
|---|---|
| Explanatory model with rules | Conceptual model with intention |
| Logic and cause/effect as concepts | Intention, sympathy, and empathy as concepts |
| Attempting to manipulate and control the external world. | Trying to understand, accept, and experience the inner world |
| Makeability as an ideal | Liveability as an ideal |
| Science, cool understanding, and calculable phenomena | Art, the warm feeling, and the incalculable phenomena |
| Rational conviction | Rhetoric seduction |
| "Measuring is knowing" | "Who measures does not get to know anything" |
| Assessment criteria as decisiveness, effectiveness, and profitability | Assessment criteria as curiosity, wisdom, and concern |
| Herrschen als Grundmotiv der Weltanschauung (Ruling as the basic motive of the world view) | Lieben als Grundmotiv der Weltanschauung (Loving as the basic motive of the world view) |

Van Kemenade (2009, 2010, 2011) integrates the commitment paradigm of Vinkenburg (2006) in the paradigms of Jouslin de Noray, described above. He calls them value orientations in line with Beck and Cowan, as described in Chap. 3.4. He comes to five value orientations that are relevant at this moment. In this context it concerns the totality of values, assumptions, suppositions and convictions regarding quality and how quality may be realized within companies. The assumptions are related to what is the function and definition of quality; what are the main quality objects; what are the basic rules for quality; who are the stakeholders; which procedures are suitable for this; and which subjects play a role in the quality assessment. Van Kemenade distinguishes five value orientations: control, continuous improvement, commitment, breakthrough, and reaching the essence.

### 4.1.1 Control (Order)

External assessment is often mandatory to enable employees to function according to this order, since the eyes of strangers are compelling?

Core values within this value orientation in terms of Beck and Cowan are *Truth* and *Order*. There are strict rules. External assessment is often mandatory enabling employees to function orderly, since the *eyes of strangers are compelling*. The management style is based on giving top-down instructions and orders according to the rules. The motto is: *"Everything has a purpose, a place and a reason"*. Following Graves (1974), Beck and Cowan (1996), and Roozendaal (2008) refers to the … of this value orientation as: the great good, the true doctrine and righteousness. Bertrand Jouslin de Noray: "The manager determines and follows the budget".

Accreditation in higher education and health care is an example of an external evaluation fitting in this paradigm. In terms of Garvin, we refer to the production-oriented approach.

In organizations operating based on the value orientation, material, and commercial assets are the key elements (Hardjono, 1995). The organization is oriented to effectiveness or efficiency. Effectiveness has the risk of compartmentalization (from a marketing myopia). Efficiency entails the risk of bureaucratization (from inertia). If we place this value orientation on the axes of Hardjono and his four-phase model (see Chap. 5) (internal versus external, control versus change), the value orientation of control fits into the quadrant control/internal, aimed at increasing efficiency.

van Schijndel and Berendsen (2007) make a distinction between organizational, professional, and relational quality. Organizational quality is expressed in the way in which the organization manages to control and improve the business processes. Professional quality relates to craftsmanship of content. An organization's relational quality is determined by the way in which employees and management relate to each other and to their customers. In our opinion, control of organizational quality is key within the control value orientation.

Crosby (1979) and his conformance to requirements fit within this value orientation. A metaphor for this value orientation is the army, or the (organization of) the Vatican.

### *4.1.2 Continuous Improvement (Success)*

Continuous improvement as a means of on-going success and performance

Beck and Cowan refer to *Enterprise* as value orientation. The core value is that an organization depends on continuous improvement as a means of on-going success and performance. The management style is doing business, based on plans, to reach objectives to benefit yourself and a number of other parties. The motto is: *"People are meant to succeed and become winners"*. Following Graves (1974), Beck and Cowan (1996), and Roozendaal (2008) describes the value structure of this value orientation as: winning, being the best, and status.

Aside from the production process, this also more and more considers the importance of other processes, such as design and marketing processes, as well as the customer's interest. The term quality *care* is quite appropriate within the continuous-improvement value orientation. Characteristic of quality care then is that the four steps of the Deming cycle are followed in order to achieve performance optimization. But there is less of an emphasis on standardized norms; less emphasis on control and inspection, often even represented in the form of a PDSA-circle: PLAN, DO, STUDY, ACT. In this context, the *check* step, showing strong associations with control, is replaced by *study*, or statistical analysis. Performance indicators are an important tool for steering continuous improvement. But also, as Betrand Jouslin de Noray says: *"The manager coaches the team to win"*.

In organizations operating based on this value orientation, the commercial assets are the key (Hardjono, 1995). When we put this value orientation on the axes of Hardjono (1995), the paradigm of continuous improvement fits in the quadrant control/external, aimed at expanding effectiveness. In terms of Hardjono, the organization is system-oriented. For the organization in its totality, models are applied such as those of the Malcolm Baldrige Award, the European Quality Award (EFQM), and the Dutch Quality Prize (the INK management model), or ISO 9001:2015. In principle, methods, such as *lean production* and Six Sigma, fit in the continuous improvement movement. By the way, Hardjono and De Klein (2004) asserts that *the EFQM model has degenerated to a controlling model* and then the EFQM model would belong to the previous value orientation. Carter and Swanwick have also adopted this thought.

In terms of Garvin (1987), it is a matter of a user-oriented approach, in view of the user's interest in the assessment of the product's quality.

According to De Schipper (2004), value orientation fits with the third, system-oriented phase of the INK management model. In our opinion, in terms of Schijndel and Berendsen (2007), the improvement of professional quality is the key element of this paradigm. Important names in this context, aside from Deming are: Feigenbaum (1986), Imai (1986) and Kaizen, Juran (1989) and his fitness for use.

A metaphor for this value orientation is the period of the Enlightenment after the Middle Ages or the present Silicon Valley.

### *4.1.3 Commitment (Community)*

There are several stakeholders, each with their own basic rules

For the value orientation of involvement and a sense of community in terms of Beck and Cowan, we are talking about *Communitarian*, the community as the core value. The management style consists of collaboration with others to reach agreements and share feelings in order to improve situations. The motto is: *"There is plenty of room for everyone"*. Following Graves (1974), Beck and Cowan (1996), and Roozendaal (2008) mentions compassion and support as the value structure of this paradigm. It fits in what McAdam-O'Connell (2005) refers to as the *visionary ethos frame*. Its characteristics are among other things: cooperation, responsibility, networks, circulating power, and diversity.

Professionals cooperate and are organized in teams, whereby the responsibility for the result is focused on each other and on fellow human beings. Yet there are more stakeholders combined (e.g. customers, personnel, government, management, suppliers, and partners) each with their own basic rules. Based on what will you let one party prevail over another? In involvement thinking, the outcome of the *dialogue with the stakeholders* is not determined in advance, nor is whose opinion will be the deciding one at a particular time.

Shoji and Walden (2006) do not distinguish a comparable paradigm. Spending an entire chapter on communities and societal values, they see a growing need for learning together and for integration with a number of societal responsibilities as a condition for the following systemic paradigm, in which renewal and innovation are core elements. The transition from continuous improvement to the following paradigm may, according to Shoji and Walden (2006), be hindered by arrogance *(what is already being made and sold is what customers will always want)*. To break this, one must look at the fundamental company objectives in order to find new values that exceed current business interests. New societal infrastructures are needed for future know-how. This goes beyond networking from the paradigm of the continuous improvement and relates to communities.

Each organization must determine their own quality description in a continuous discussion with various stakeholders. Contrary to the previous views of quality, that may be called rationalistic, with this paradigm fitting the hermeneutical approach. It concerns *excellence through passion.*

Organizations in this view, according to the terminology of Hardjono (1995), have a strongly developed capability for socialization. They are oriented toward flexibility which has a risk of anarchy (from chaos). When we put this value orientation on the axes of Hardjono (1995), the paradigm of involvement fits in the quadrant change/internal, aimed at expanding flexibility. According to De Schipper (2004), the value orientation fits with the fourth, chain-oriented phase of the INK management model.

Where Carter and Swanwick (2006) have questions regarding the defensibility of quality models such as the EFQM/European excellence model, they propose two alternatives: contextualized models (each organization makes its own model) or human excellence (individual quality models). Both solutions have in common with each other that it concerns the boosting of the organization's or individual's intrinsic motivation. Both alternatives fit in the value orientation of involvement and communality. In higher education being accountable through dialogue with the help of Audit Instrument for Sustainability in Higher Education (AISHE, see www.dho.nl/aishe) is perhaps a development that fits within this value orientation (Wolf, 2015).

In our opinion, in terms of van Schijndel and Berendsen (2007), it is not the organizational nor the professional, but the relational quality that is at the core.

A metaphor for this value orientation is the family, the commune from the sixties of the previous century, Woodstock.

### 4.1.4 Breakthrough

> The organization is a systematic flow of which the structure agrees with the momentary task.

Beck and Cowan (1996) distinguish two levels in their spiral dynamics. After the value system of *community* follows the value system with *synergy* at the core. They mention this transition as a transition to another level (Graves referred to it as second-tier thinking). The management style is supporting people to provide know-how, information and resources and to pave the way (release), so people may do their job. The motto is: *"If it cannot be done the way it should, then it should be done the way it can"*. Tolerance is a key term. The organization is a systematic flow, the structure of which agrees with the momentary task. The organization may adopt other structures if the context requires it. This manner of thinking is also found in the theory of living systems and cybernetics, but also the work of Rawls and Rorty. Following Graves, Beck and Cowan, and Roozendaal (2008) mentions understanding, validated know-how, and dialogue of the value structure of this paradigm.

Organizations in this view, according to the terminology of Hardjono (1995), have a strongly developed intellectual capability. They are oriented toward creativity and this has the risk of hobbyism (from an oversensitivity for impulses from the environment). When we put this value orientation on the axes of Hardjono (1995), the value orientation of breakthrough fits into the four-phase model (see Chap. 5), quadrant change/external, aimed at expanding the flexibility. According to De Schipper (2004), this value orientation fits into the fifth phase of the INK management model, which people referred to as *transformation.*

The breakthrough of radical innovation may express itself in societal change: the acceptance of a particular role in society by means of CSR-based entrepreneurship. Corporate social responsibility can be mentioned in each of the five value orientations, but with a totally different content and meaning.[2] From this value orientation can one really speak of CSR.

Particularly, Jouslin de Noray as well as Shiba have focused on deepening the breakthrough-paradigm. Bertrand Jouslin de Noray: *"The manager shows the star"*. A metaphor for this value orientation is system thinking, the network organization.

### 4.1.5 Reaching the Essence

The manager puts spiritual values on the forefront providing freedom and protection. He himself is inhabited by these values.

Beck and Cowan (1996) subsequently distinguished the system of values centering around a system of holistic living. It is the phase in which the world is seen as a totality that is continuously evolving. The management style is based on working with global and local networks to address the long-term problems affecting all forms of life. The motto is: *"Everything is connected to everything"*. According to Jouslin de Noray, the dominant behavioral pattern in this philosophical approach is that people in organizations want to understand the question regarding the essence of matters.

---

[2]See European Corporate Sustainability Framework: http://www.sustainability-reports.com/titel-652/.

Mutual dependence is the key term. The concept of organization in the form of network organizations fits in with this (Whetten, 1981; Powell, 1990). Following Graves, Beck and Cowan, and Roozendaal (2008) mentions understanding, validated know-how, and dialogue of the value structure of this paradigm. Willingness to surrender profit is in favor of sustainability. Jouslin de Noray: *"The manager puts spiritual values on the forefront providing freedom and protection. He himself is inhabited by these values"*.

In his Dutch publication *Verdraaide organisaties* (twisted organizations), Hart (2012) argues that organizational concepts that are driven by the world of systems, via the living environment to the purpose, should be turned around to be driven from purpose to living environment to the world of systems. Earlier, back in 2009, Sinek presented his *Golden Circle* in which he stimulates leaders to no longer think from "what", "how" to "why", yet from "why" to "how" and "what".

Zohar (2004) speaks of a spiritual capability of organizations. Following Hardjono (1995), one might state that these organizations have a strongly developed spiritual capability.[3] They are oriented toward the ultimate purpose. This entails the risk of haziness (from striving for the superhuman) or rigidity (from striving for the ultimate truth). Moreover, the concept *spiritual capability* does not exist in the four-phase model (Chap. 5). Hardjono acknowledges the existence of this capability. Based on a suggestion by Bishop Van Luyn, Hardjono refers to it as *pneu*: the ability to maintain contact with the supernatural.

Writers in various religious and spiritual movements have considered entrepreneurship. In 1999, De Vries, who specialized in research into ISO, authored the booklet *Kwaliteitszorg zonder onbehagen* (quality care without discontent), with the sub-title of (literally translated): "Practical advice for the use of ISO 9000 as the outcome of a Christian philosophical analysis". In this, he describes how dissatisfaction about ISO 9000 may be attributed to two conflicting ideals: the freedom ideal and the control ideal. He seeks to find a way out in ecumenical-Christian philosophy: *Quality care has to do with the deepest motivations both of individuals as well as of the company in totality. Individually: to which degree is their freedom concept a striving for autonomy? And collectively: resistance may be solved when individuals meet each other "as a human being"* (Geelhoed, 1996, p. 45)".

Together with Howard Cutler, the Dalai Lama wrote an entire book about the relation between happiness and work: The Art of Happiness at Work (Dalai Lama and Cutler 2004). Quotation: *"I think that it is important not to forget that all human activities, work or other, mainly must be adressed to increase the wellbeing of mankind" and: "One starts with the full awareness of the importance of the basic human values, which, after my opinion, are the source of true happiness, at work and and home. Values as kindness, tolerance, compassion, honesty and forgiveness. "Working according the Budhistic principles is then living according the Budistic principles. One lives, peels patatos, coocks, does the dishes and works in full awareness. From moment to moment living life in conscious attention. One has to be busy with what*

[3]See also: http://www.degoudseschool.nl/new-blog/2016/7/1/7q2c3l52v2ntmwefwmvg285gvwhis9.

*one is doing here and now and not busy with the past nor the future: 'the work from now'.*

Derkse (2003) has influenced leaders with his *Life Principle for Beginners. Benedictine Spirituality for Daily Life.* Alsalmani (2017) tried to find a quality management system fitting with Islamic business ethics. However, there is a major risk that these faith-inspired rules will be applied too much as top-down instructions, which brings us back to control and the Empiric Paradigm.

Foundation De Goudse School,[4] established by Hardjono and Roozendaal, fits this system of values. They want to distinguish themselves by finding the link between spirituality and technological developments. In this way, they want to provide a contribution to the discussion where society should be heading, which values must also apply for the future, and what this eventually means for the way society is organized as well as for the organizations operating in it.

A metaphor for this value orientation is intuitive thinking, living organizations, and global networks with a global objective.

These value orientations are summarized in Table 4.2. We are of the opinion that the manners of thinking about quality do not 1-to-1 relate to value orientations. In Chap. 9, after the description of paradigms, we will discuss the relationship between paradigm and value orientations.

## 4.2 Schools and Value Orientations are of a Different Order

The value orientations of Beck and Cowan and the movements of Jouslin de Noray and Shiba can easily be linked to each other. In the value orientations of van Kemenade (Sect. 4.1), the link is established. The schools and the classification of van Kemenade initially appear to be overlapping one another. However, each of the value orientations fits, sometimes to a lesser degree, sometimes to a greater degree, in different schools. In our opinion, the schools as named by Vinkenburg are of a different order. But we see no reason why not one of the schools could be the starting point of one of the movements as named in Van Kemenade (2010). So, it could be a case of reflective thinking in the movement of *control*, even though this could introduce tensions, other than reflective thinking does with "reaching the essence".

Visually, the value orientations and schools might be placed on a horizontal and vertical axis.

According to us, a school is a movement with members, sharing the same opinions about the entity to which they are oriented, and sharing the same opinion on how to behave themselves, what they want to focus on, and with which intention. A school may have the characteristics of a paradigm, that is participants do not recognize the other schools or reject their principles. However, each school has its limitations that cannot be solved within that particular school. Within a school, it is possible that multiple value orientations are recognized, such as improvement, involvement,

[4]http://www.degoudseschool.nl/new-blog/2016/5/25/degoudseschool.

**Table 4.2** Five value orientations according to Van Kemenade (2010)

| Function/intended operation | 1. Control | 2. Continuous improvement | 3. Involvement | 4. Breakthrough | 5. Reaching the essence |
|---|---|---|---|---|---|
| Value orientation Beck and Cowan | *Truth/order (Blue)* | *Enterprise/Success, (Orange)* | *Communitarian/Community (Green)* | *Synergy (Yellow)* | *Holistic life system (Turquoise/Teal)* |
| Quality = | Conformance to requirements<br>The degree to which a totality of properties and characteristics meets the requirements | Fitness for use<br>Meet or exceed customer expectations as well as continuous improvement | Fitness for purpose<br>The degree in which the intentions of all stakeholders are met, in consideration of here and now, there and the future | Quality is transformation<br>Contribution to actual innovation (ECSF, 2004) | Quality is transcendental<br>Contribution to a better world (Vinkenburg, 2006) |
| Stakeholder | The customer, the supplier | The customers and partners | All stakeholders | All stakeholders now, here and then, there | |
| Management style | Top-down instructions and giving orders according to the rules | Doing business, based on plans, to reach objectives to benefit yourself and a number of other parties | Collaboration with others to reach agreements and share feelings to improve situations. | Supporting people to provide know-how, information and resources and to pave the way (release), so people may do their job | Working with global and local networks to address the problems affecting all forms of life in the long term. |
| Motto | "Everything has a purpose, a place, and a reason" | "People are meant to succeed and become winners" | "Plenty of room for everyone" | "Open to learning at any time and from any source" | "Everything is connected to everything" |
| Responsibility | *Vertical responsibility* | | *Horizontal responsibility* | | |

(continued)

**Table 4.2** (continued)

| Function/intended operation | 1. Control | 2. Continuous improvement | 3. Involvement | 4. Breakthrough | 5. Reaching the essence |
|---|---|---|---|---|---|
| View of corporate social responsibility (CSR) Hardonjo et al. () | CSR at this level consists of providing welfare to society, within the limits of regulations from the rightful authorities. In addition, organizations might respond to charity and stewardship considerations. The motivation for CSR is that CSR is perceived as a duty and obligation, or correct behavior | CSR at this level consists of the integration of social, ethical, and ecological aspects into business operations and decision-making, provided it contributes to the financial bottom line. The motivation for CSR is a business case: CSR is promoted if profitable, for example because of an improved reputation in various markets (customers/employees/shareholders) | CSR consists of balancing economic, social, and ecological concerns, which are all three important in themselves. CSR initiatives go beyond legal compliance and beyond profit considerations. The motivation for CSR is that human potential, social responsibility and care for the planet are important as such | CSR consists of a search for well-balanced, functional solutions creating value in the economic, social, and ecological realms of corporate performance, in a synergistic, win-together approach with all relevant stakeholders. The motivation for CSR is that sustainability is important in itself, especially because it is recognized as being the inevitable direction progress takes | CSR is fully integrated and embedded in every aspect of the organization, aimed at contributing to the quality and continuation of life of every being and entity, now and in the future. The motivation for CSR that sustainability is the only alternative since all beings and phenomena are mutually interdependent. Each person or organization, therefore, has a universal responsibility toward all other beings |
| Manner of thinking | Modern/Anglo-Saxon | Modern/Japanese | Modern/Mediterranean | Modern/Rhinelandic | Post-modern |

(continued)

**Table 4.2** (continued)

| Function | 1. Control | 2. Continuous improvement | 3. Involvement | 4. Breakthrough | 5. Reaching the essence |
|---|---|---|---|---|---|
| Names | Taylor, Crosby | Juran, Kano, Deming, Imai, Hardjono (1995) | Vinkenburg | Jouslin de Noray, Shiba | Zohar, Hardjono |
| Jouslin de Noray | Process control | Integral QM | – | Breakthrough | Reaching the essential |
| **Shiba**<br>Change<br>Unchange<br>Human<br>Key player<br>Management focus | Process<br>Standards<br>Theory X<br>Middle manager<br>Discipline focus | Standards<br>Business<br>Theory Y<br>Shop floor workers<br>Backward focus | Relations<br>Business<br>Theory Y<br>All employees<br>People focus | Business<br>Values<br>Theory Z<br>Top managers<br>Forward | Values<br>All stakeholders<br>From why to how to what |
| Hardjono<br>Generation model<br>Phase characteristic<br>Expanding of assets<br>Risk | Generation 1 (product-oriented) and 2 (process-oriented)<br>Oriented toward efficiency<br>Material assets<br>Bureaucratization (from inertia) | Generation 3 (System-oriented)<br>Orientation toward effectiveness<br>Commercial capability<br>Compartmentalization (from myopia) | Generation 4 (Chain-oriented)<br>Orientation toward flexibility<br>Socialization capability<br>Anarchy (from chaos) | Generation 5 (Society-oriented)<br>Orientation toward creativity<br>Intellectual capability<br>Hobbyism | -<br>(Orientation toward intention)<br>Spiritual capability<br>Haziness (from striving for the superhuman) or rigidity (from striving for the ultimate truth). |
| Role of leader | *Calculating* | *Negotiating/selling* | *Coaching* | *Participating* | *Delegating* |
| Role of Quality manager | Inspector | Advisor | Facilitator | Builder of bridges | Change agent |

(continued)

**Table 4.2** (continued)

| Function | 1. Control | 2. Continuous improvement | 3. Involvement | 4. Breakthrough | 5. Reaching the essence |
|---|---|---|---|---|---|
| Competencies of Quality manager | Statistics, accuracy, auditing, technological skills | Professional skills, performance management, diplomacy, persuasive power, promotion of customer orientedness | Providing support, relational skills | Collaborating, promoting synergy | Change science |
| Context | There is a fire, or war | There is a great competition | You cannot do it alone internally | You need assistance from external organizations | We look for meaning |
| Metaphor | Army, Vatican | The Enlightenment, Silicon Valley | Family, Commune (Woodstock) | Systematic thinking, networking | Living organizations, thinking intuitively, global networks for global objectives |

breakthrough, or reaching the essence, and that adherents know how to act based on this; also with the possibility that one rejects one of these items. In practice, people may need to apply another school—depending on a particular context. A proper mix may prevent that one of the schools will drift into extremes leading to the opposite of what was intended. To us, it is definitely a matter of paradigms, even when you accept that someone can make use of several paradigms. Just as with a school it is in our opinion possible to apply the movements or phases within a single paradigm, as De Noray calls them: "control", "continuous improvement", "breakthrough" and "reaching the essence". Vinkenburg (2006) has the opinion that the schools distinguish themselves by the object on which their attention is focused: the empirical school focuses on the product/thc process; the normative school focuses on the organization/the system; the reflective school focuses on the group/the human being. Also, in the normative school we observe much attention for the product and the process. Even systems/organizations and even groups/human beings may be approached in an empirical manner.

It is not the object that determines the school

We are of the opinion that the schools may study any object, but that it concerns different way of observation. In SqEME® terms it concerns four different windows with which you may approach any object. So we obviously do not share Vinkenburg's opinion that the object determines the school.

## 4.3 We are Searching for Paradigms

A paradigm is a specific collection of questions, viewpoints and models

We here deliberately decide to talk about paradigms in thinking about quality and, for example, not for quality-science-related schools or value orientations. There are too many definitions of what is a paradigm. Traditionally, authors quote Kuhn (1962). Kuhn proposed that a paradigm defines *"the practices that define a scientific discipline at certain point in time"*. He states that the paradigm: *"universally recognized scientific achievements that–for a time–provide models, problems, and solutions for a community of practitioners"* (Kuhn, 1962). Thereby, he speaks particularly about paradigms in science.

A paradigm should describe:

- What is to be observed
- The kind of questions that are supposed to be asked and probed for answers in relation to this subject
- How these questions are to be structured
- What predications are made by the primary theory within the discipline
- How the results of scientific investigation should be interpreted

- How an experiment is to be conducted and what equipment is available to conduct the experiment.

We follow the simpler, more popular definition, derived from Salter and Wolfe (1990): *"A paradigm is a specific collection of questions, viewpoints and models that define how authors, publishers and theorists who subscribe to that paradigm, view and approach the science"*. In our case, this concerns the collection of questions, views, and models determining the quality.

A paradigm meets the rule of incommensurability, that is in principle the paradigms are incompatible with each other. *"Paradigms in Kuhn's original intention are by definition incommensurable and without the latter notion the paradigm concept largely loses its scientific theoretical impact"* (Essers, 2007). We endorse that paradigms are mutually exclusive. However, we believe that a quality expert or an organizational expert should be able to look beyond paradigms and combine what is needed at a given moment in a given place. So, a quality expert/organizational expert must *be able to combine* methods from various paradigms, mentioned above as epistemic fluency.

## 4.4 Conclusions

In this chapter, we investigate what others have stated about movements, value orientations, schools of quality. We compared these and have attempted to find the link. Are these schools, as Vinkenburg (2006 and following) proposes? How do these manners of thinking relate to the revolutions of Jouslin de Noray, Shiba, to the value orientations of Beck and Cowan?

We come to the following conclusions:

1. We have attempted to integrate the thoughts of Jouslin de Noray, Shiba, Beck and Cowan (and Vinkenburg's involvement paradigm) with each other in five value orientations that are interesting for quality management.
2. The manners of thinking about quality, which we will develop in more detail in the following chapter, may definitely be referred to as paradigms in the definition of Salter and Wolfe (1990). We do not opt for the *school* concept, nor we opt for value orientations.
3. Vinkenburg's schools cannot be linked 1-to-1 with one of the value orientations. You may engage in quality management/control based on the Empirical Paradigm, the Reference Paradigm, or the Reflective Paradigm. Value orientations are of different order compared to paradigms. Based on each paradigm, you can work on one of the value orientations even though it will result in more friction with the one than with the other.
4. *It is not the object that determines the school.*
5. *A paradigm is a specific collection of questions, viewpoints, and models.*
6. We must continue our quest for the fourth paradigm.

# References


Al-Salmani, N. (2017). Quality management guidelines for Islamic societies. Amsterdam: VUPress.

Beck, D., & Cowan, C. (1996). *Spiral dynamics*. Malden: Blackwell Publishers.

Crosby, P. B. (1979). *Quality is free*. New York: McGraw-Hill.

Derkse, W. (2003). *Levensregel voor beginners*. Lannoo, Tielt: Benedictijnse spritualiteit voor het dagelijkse leven.

De Schipper, I. J. P. J. (2004). *NK-ontwikkelingsfasen en de rol van waarden. Verkenning van de positivistische en interpretatieve benaderingen voor het beschrijven van organisaties in ontwikkeling*, Eindhoven.

Essers, J. P. J. M. (2007). *Incommensurabiliteit en organisatie: De reconstructie van een academische patstelling* (p. 9789058921338). ISBN: Thesis for obtaining the doctoral degree of RSM Erasmus University Rotterdam.

Feigenbaum, A. V. (1986). *Total quality control*. New York: McGraw-Hill.

Geelhoed, M. J. (1996). *Cultuur en normativiteit—Kleine wijsgerige beschouwing over weerstand tegen verandering en cultuur, doctoral thesis for culture, organization and management*. Amsterdam: Free University.

Garvin, D. A. (1987). Competing on the eight dimensions of quality. Harvard Business Review. Retrieved December 20, 2016.

Graves, G.M., & Geoffrion, A. M. (1974). Multicommodity distribution system design by Benders decomposition. *Management Science, 20*(5), 822–844.

Hardjono, T. W. & De Klein, P. (2004). Introduction on the European Corporate Sustainability Framework (ECSF). *Journal of Business Ethics, 55*(2), 99–113.

Hardjono, T. W. (1995). *Ritmiek en organisatiedynamiek*. The Hague, Kluwer: Vierfasenmodel.

Imai, M. (1986). *Kaizen: The key to Japan's competitive success*. New York: McGraw-Hill Education.

Juran, J. M. (1989). *Juran on leadership for quality, an executive handbook*. New York: The Free Press.

Hart, W. (2012). *Verdraaide organisaties*. Deventer: Kluwer. ISBN 9789013105735.

Kuhn, T. S. (1962). *The structure of scientific revolutions*. Chicago/London: The University of Chicago Press. ISBN 0226458083.

Lama, D., & Cutler, H. C. (2004). The art of happiness at work by Dalai Lama. Hodder Mobius. ASIN: B01K3PYNSQ ISBN-10: 9780340750155.

McAdam-O'Connell, B. (2005). Changing conversations in the process of transformational change: European Quality Movement 1998–2004, Department of Sociology National University of Ireland Cork.

Powell, W. W. (1990). Neither market not hierarchy: Network forms of organization. *Research in Organizational Behavior, 12*, 295–33.

Roozendaal, A. (2008). *Contextueel leiderschap*. Van Gorcum.

Salter, L., & Wolfe, D. (1990). *Managing Technology: A Social Science Perspective*. Toronto: Garamond.

Shoji, S., & Walden, D. (2006). Breakthrough management: Principles, skills, and patterns or transformational leadership, confederation of Indian industry. ISBN-10: 8190356437, ISBN-13: 978-8190356435.

van Kemenade, E. A. (2009). *Certificering, accreditatie en de professional*. Eburon uitgeverij.

van Kemenade, E. A. (2010). Past is prologue. Know the history of quality management to achieve future success. *Quality Progress*. Retrieved 4th of March from https://www.researchgate.net/publication/259819431_Past_is_Prologue_Know_the_history_of_quality_management_to_achieve_future_success.

van Kemenade, E. (2011). Briefwisseling over scholen in kwaliteit. *Sigma* 5. oktober 2011. Retrieved 4th of March from https://www.vankemenade-act.nl/wp-content/uploads/2017/08/SIGMA05_Briefwisseling.pdf.

van Schijndel, B., & Berendsen, G. (2007). Kwaliteit is mensenwerk; relationele kwaliteit als kwaliteitsfactor. *Synaps* (23), 7–11.

Vinkenburg, H. (2006). Dienstverlening; paradigma's, deugden en dilemma's. In *Kwaliteit in Praktijk B1-5.*

Whetten, D.A. (1981). Interorganizational relations: A review of the field. *Journal of higher education.* Vol. 52, Nr.1.

Wolf de, M. (2015). In dialoog naar een kwaliteitscultuur in het HBO. *Sigma* 6. Retrieved, 4th June 2017 from http://www.sigmaonline.nl/2016/05/in-dialoog-naar-een-kwaliteitscultuur-in-het-hbo/.

Zohar, D. (2004). *Spirituele waarde.* Utrecht: Kosmos Z&K publishers.

# Chapter 5
# Thinking in Four

We may perhaps be able to draw material from existing divisions in fours in our quest for the fourth paradigm. This paradigm more fully explains past developments, such as the Japanese approach to quality and recent developments in health care, and it is a paradigm tailored to thinkers such as Deming and Pirsig. We start the search from what we know of the above-mentioned three paradigms. The number 4 may be a *holy* number, but to us that does not constitute a reason to look for a fourth paradigm. In spite of the fact that the number is associated with earthly matters and that there are 4 seasons, 4 winds, 4 evangelists, 4 elements, and so on. The number 4 also symbolizes the virgin Mary. There are four fundamental forces of nature: The strong core force which keeps things together, the electromagnetic force, the forces between particles, the weak core force playing a significant role in processes of decay, and the force of gravity which pulls things together.

More important is the fact that much research, many researchers and thinkers, particularly in business, eventually find the two-by-two matrix, the so-called Harvard Matrix. The Harvard Matrix enables us to compare the views and discoveries and to transpose the supplied evidence. Moreover, such a matrix functions similar to the *periodic system* in chemistry. It does not merely order the elements; it also helps to find gaps and allows one to utilize the same characteristics for the description of elements.

In fact, there are two types of $2 \times 2$ matrices. The first is based on two axes whereby a point moves along the axes of the matrix or may be positioned on it. The matrix may consist of several grids $3 \times 3$, $4 \times 4$, and so on. Or even a third axis or dimension may be added to it. The matrix then becomes a cube. The other type is a division that gives meaning. Then it is no longer a matter of different axes, each with its own dimension, but of directions of orientation. An idea is then put in the matrix where it agrees with the combination of its included orientation directions. Sometimes the diagram pictures the four cells or categories independently of one another. The SqEME© model later to be described is one example of this. SqEME© discusses four different windows, in which each window provides a different image



T. Hardjono and E. van Kemenade, *The Emergence Paradigm in Quality Management*, https://doi.org/10.1007/978-3-030-58096-4_5

of the object. In order to compare matrices with each other, it may be necessary to rotate them with regard to the way they are usually displayed. We will definitely do this in a particular case.

## 5.1 Wilber's All Quadrant Model

Wilber not only provides a description for the three paradigms, he also directs the way towards a fourth paradigm.

Wilber (2000) at the start of his book *The Theory of Everything* mentions four aspects: individual/interior; individual/exterior; collective/interior; and collective/exterior (see Fig. 5.1). Initially, Vinkenburg detects three schools in quality care in Wilber's theory: the empirical school, the normative school, and the reflective school. However, later he abandons Wilber's views.

Kemenade and Hardjono (2011) felt that Wilber not only provides an indication for the three paradigms, but that he also refers to a fourth, in which collective values play an important role (see Fig. 5.2).

In the scientific world, Wilber is exposed to extensive criticism. He has the tendency to quite assertively deal with serious scientists criticizing his *Theory of Everything* (see, e.g. http://www.kenwilber.com/blog/show/46). Yet, he does make you think. In our book, we still have adopted Wilber's ideas.

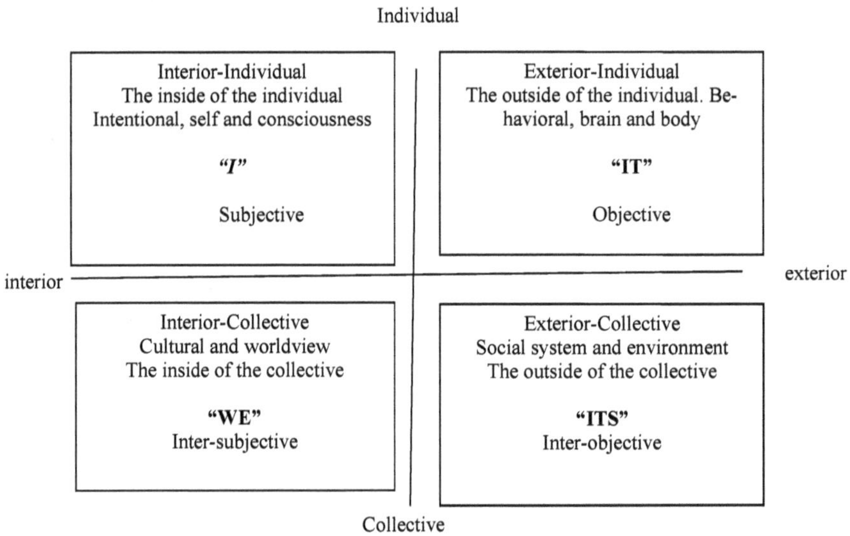


**Fig. 5.1** Four Quadrants by Wilber (2000)

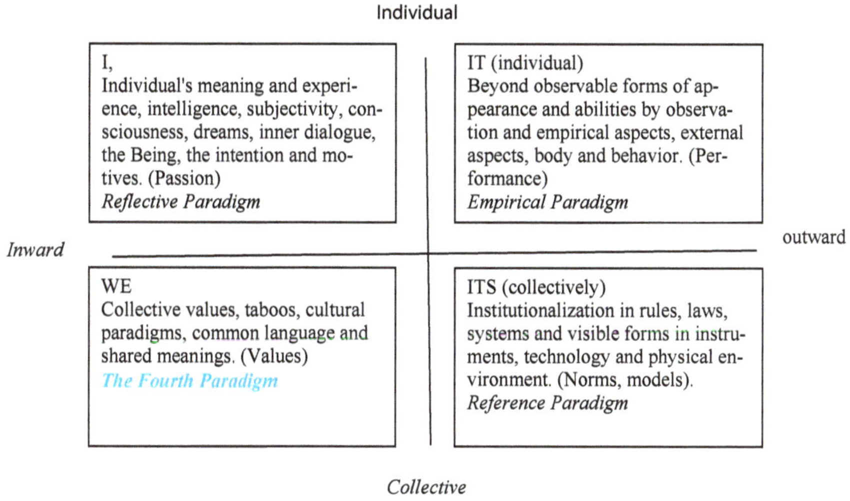


**Fig. 5.2** The four quadrants according to Wilber adapted by Van Kemenade and Hardjono, 2011)

## 5.2 Hardjono's Four-Phase Model

Also Quinn's division, as Hardjono has applied it in his four-phase model, refers to a fourth paradigm. Hardjono (1995) in his four-phase model mentions the organization as an object in his thesis *Rhythmics and Organizational Dynamics*. According to Hardjono, an organization's objective is the enlargement of capabilities, derived from Pirsig's values hierarchy. Although, according to Hardjono, the hierarchy is determined by a decisive strategic orientation within a particular timeframe. He lists the capabilities as follows: (see also Fig. 5.3).

1. The material capabilties or competences of an organization reported on the balance sheet as such or the tangible equity;
2. Commercial capabilities or competences, the ability to enter into transactions with third parties;
3. Socialization capabilities or competences, the capability to evolve as a group and to act as a group;
4. Intellectual capability or competence to think, the collective result of personal growth, and intellectual development of the members of the organization.

The four-phase model, based on two dichotomies, does not understand the one if the other is unknown, even though they are opposites of each other. Control driven is juxtaposed to change driven. External driven, allowing yourself to be controlled by the environment, is opposite to allowing yourself to be controlled by your own values, internal driven (Fig. 5.3).

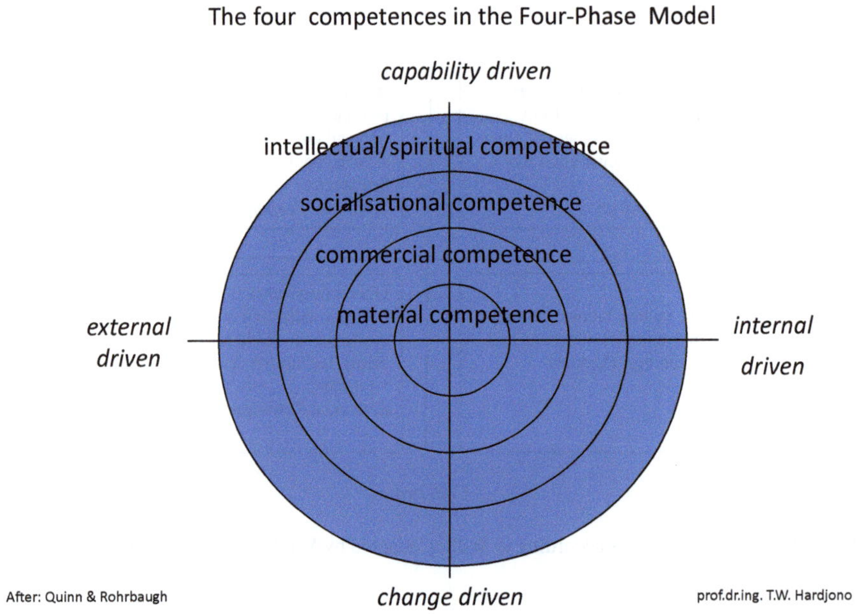


**Fig. 5.3** The four capabilities or competences in the four-phase model

Organizations must select the most dominant approach. The four-phase model calls these strategic principles: they include *effectiveness, efficiency, creativity, and flexibility*.

Banning waste, *efficiency*, may be regarded as a result of control-driven and internal-driven, see Fig. 5.4. The outcome must be empirically proven, and the willingness to accept hierarchy/authority is at its core. Diagonally opposed is change-driven. Showing a willingness to change and focusing on the environment is external-driven, a phase in which the limitations are always expanding, while thinking laterally, whereas people are open toward emergence more than in other phases.

The quadrant indicated as *effectiveness* within the four-phase model is characterized by the willingness to adapt oneself to accommodate public opinion, that is external-driven. The willingness to come to agreements is at the core of this. Now in combination control-driven matters.

Diametrically opposed to this is change-driven, characterized by internal-driven with the willingness to change. The value-based approach comprises the core.

Effectiveness is the measure in which an organization is able to reach its objectives. Yet eventually, it may lead to compartmentalization. Efficiency is the measure in which the real effort agrees with the planned or theoretical effort, but eventually may lead to bureaucracy. Flexibility is defined as being open for, and adequately responding to the ever-changing circumstances. Yet, eventually this may lead to anarchy. Creativity is defined as the measure in which the organization is able to

*Control driven*

| *External driven* | | *Internal driven* |
|---|---|---|
| Effectiveness<br>Seeking direction<br>Making agreements<br>Danger of compartmentalization<br>*Reflective Paradigm* | Efficiency<br>Hierarchy-based<br>Striving for synergy<br>Danger of bureaucracy<br>*Empirical Paradigm* | |
| Creativity<br>Risk-taking<br>Lateral thinking<br>Danger of hobbyism<br>*The Fourth Paradigm* | Flexibility<br>Value-based<br>Realization of shared mission<br>Danger of anarchy<br>*Reference Paradigm* | |

*Change driven*

**Fig. 5.4** The four paradigms according to the Four-Phase model of Hardjono

generate new ideas and solutions in response to changing needs. Yet, eventually this may lead to hobbyism (see Fig. 5.5).

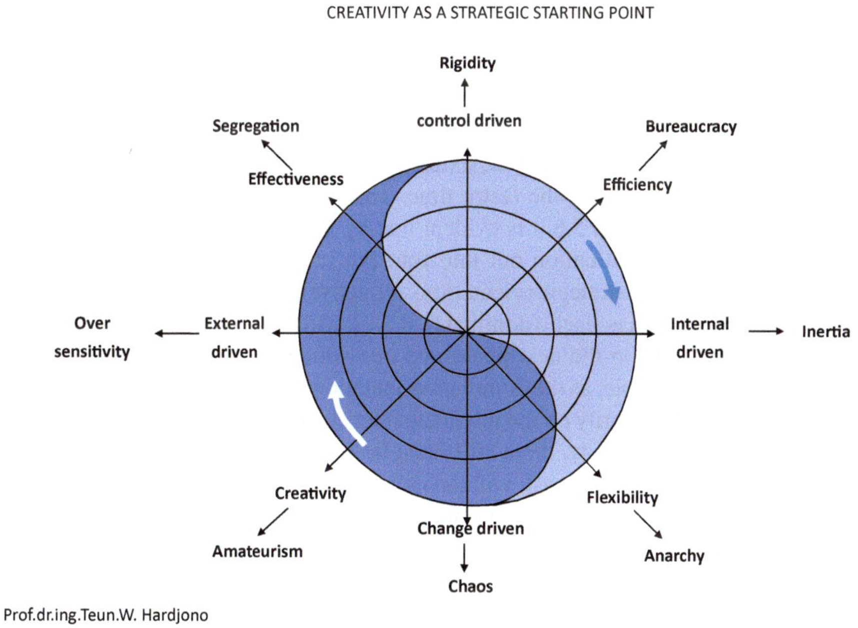


**Fig. 5.5** The Four-Phase model with creativity as a strategic starting oint

| | | |
|---|---|---|
| ***Principles*** | ***CONSTITUTION*** (Blueprint)<br>View of:<br>Philosophical principles<br>Human images<br>Operational model<br>Vision, mission, policy & strategy | ***CHEMISTRY***(Messaging traffic)<br>View of:<br>Energy exchange<br>Mutual influence<br>Communication forms<br>Motivating factors<br>*(The Fourth Paradigm)* |
| ***Rules*** | **CORRESPONDENCE** (Desktop)<br>View of: Control Systems<br>Performance Indicators<br>Feedback<br>Reward Systems<br>*(Empirical Paradigm)* | **CONSTRUCTION** (Specifications)<br>View of: Structures<br>Players<br>Stakeholders<br>Facilities<br>*(Reference Paradigm)* |

**Fig. 5.6** The windows of SqEME®

The four-phase model urges us to always think and act from a different (strategic) perspective, without losing focus on the other aspects. The four-phase model also lists sixteen organizational interventions, at every capability level by referring to an intervention belonging to a focus or orientation. According to the four-phase model, eight interventions always deserve prominent attention, while the other interventions take a background role.

In other words, the four-phase model advocates a holistic view, in which the four capabilities are always linked to sixteen interventions. Combinations of always two interventions may be connected to sixteen results in total. The four-phase model also refers to phases, that is the factor time. This is the time in terms of *chronos* as well as *kairos*. From Fig. 5.6 it is evident that by means of interventions Hardjono divides the dominant attention not into four but in fact into two. The four-phase model does show four strategic orientations (effectiveness, efficiency, flexibility, and creativity) but connects each (dominant) drives with two others. One which will be the dominant one in the next phase and one which was earlier is the dominant orientation. For example, as the dominant orientation is on creativity, there also must be, but partly and primarily on the intellectual level, orientation on effectiveness, the next phase. Simultaneously, there still must be some orientation on flexibility, the dominant orientation in the past, with an emphasis on the socialization competence The four-phase model is particularly a way of viewing things, characterized by a new fourth paradigm.

## 5.3 The Four Windows of SqEME®

We must look through all four windows to gain a complete overview.

The aforementioned divisions into four have the disadvantage of a particular rigidity, even by their structuring in two dichotomies. *SqEME®* applies a different form of division and relates to a developed method intended to gain insight into the processes within an organization. The method is applied to study an organization in its social, cultural, technological, and economical context. This relates to four *windows* constituting a description of reality. Together, the four views from these windows provide a deeper insight in this reality. The *paradigm* concept is not utilized in the *SqEME®* method. Yet, the description of what is intended with *window* comes quite close to it. The *SqEME®* method asserts that in order to gain a complete picture one has to look *through all four windows*. If a *window* were equal to a paradigm then it would be a separate reality. *SqEME®* distinguishes between two levels of abstraction. A more or less tangible/instrumental level consisting of four aspects: Blueprint, Messaging Traffic, Desktop, and Specifications, which may be a tool in describing the processes on a more or less abstract/conceptual level: Constitution, Chemistry, Correspondence, and Construct (Van Velzen et al., 2002; van Oosten, 2008). It concerns principles in the windows *Constitution* and *Chemistry*, whereas the windows *Correspondence* and C*onstruction* relate to rules. The four windows of *SqEME®* are easily linked to the four paradigms (Fig. 5.6).

## 5.4 Horizontal Organization According to Bakker and Hardjono

Yet another perspective on thinking in four aspects, also without dichotomies, may be found in the book *Horizontaal organiseren* (Bakker & Hardjono, 2013) (Fig. 5.7). They postulate that an organization *thinks* horizontally when:

- The organization is aware of existing within the social context of a network society, information society, emancipated employees, and a participation society, causing all kinds of forms of collaboration that nonetheless exists and remain independent of one another.
- The organization can handle self-control and coordination without a center, offering room for one's own responsibility placed low within the organization as well as affecting individual process performance;
- Employees understand the importance of process flows, product quality, cost, and flexibility so that delivering added value to all stakeholders is treated as the highest goal.
- Employees are able to adjust their personal goal to the organizational goal so that organizational performance is driven by collaboration and team performance.

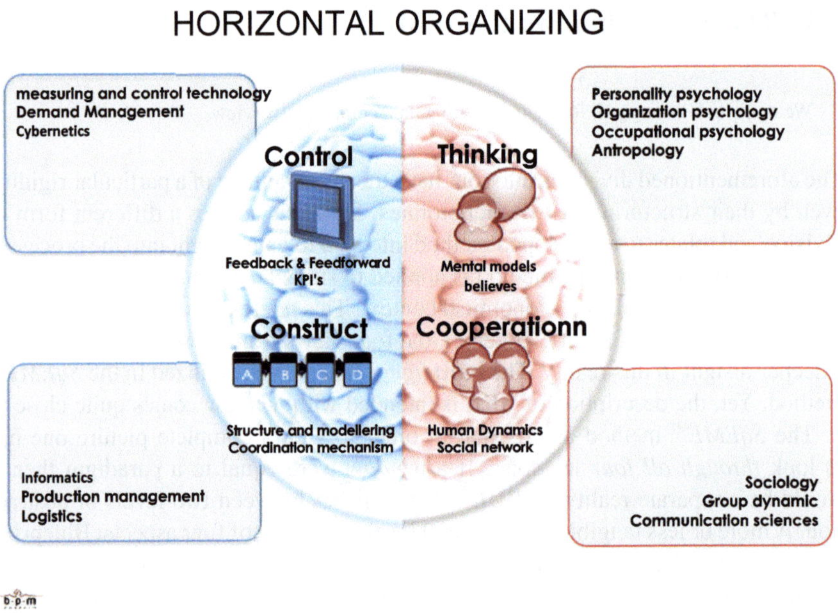


**Fig. 5.7** Horizontally organizing according to Bakker & Hardjono (2013)

The concept that an organization is *horizontally constructed* is evident from the fact that:

- The relevant organizational processes have been identified and modeled, making process flows transparent and accessible.
- Processes are designed on the basis of streamlining product quality, costs, and flexibility so that cycle time and interim adjustment are optimally facilitated;
- Processes are derived from policy and strategy so that direction, design, performance, and improvement show consistency;
- Process, personnel, and information management go hand-in-hand, making processes *empowered*.

The concept that an organization is *horizontally controlled* is evident from the fact that:

- The organizational structure is based on the process operation and the opportunity for monitoring and adjustment:
- The relevance of the process indicators is determined by the strategy;
- A balance is struck between individual freedom and by strategy-driven conditions;
- The leadership style is based on professional maturity, appreciation for the contribution made and support for continuous improvement in every conceivable way.

The concept that an organization is *horizontally collaborating* is evident from the fact that:

- Processes are approached as a social network
- Process performance is seen as team performance, with the goal of making maximum use of the collective power of thought;
- Processes are regarded as the outcome of human dynamics, in which a human being may shine as a professional, allowing the organization's performance to optimally flourish;
- The accent may be shifted over time from an emphasis on creativity, to an emphasis on effectiveness, to an emphasis on efficiency, to an emphasis on flexibility, to an emphasis on creativity, and so on.

Bakker and Hardjono thus distinguish *thinking and collaborating*, which they associate with the right hemisphere and where the visual-spatial functions are concentrated, as well as *controlling and constructing*, which they associate with the left hemisphere, where functions such as language and logic are dominant. Just as with *SqEME®* construction may be related to the Empirical Paradigm and controlling to the Reference Paradigm, thinking may be related to the Reflective Paradigm, and collaboration with the Fourth Paradigm.

## 5.5 The Cynefin Framework

The following model without dichotomies that may assist us in discovering the fourth paradigm is the Cynefin framework. It would go beyond the scope of this book to describe it in great detail. We see overlaps which we will use in more detail in Chap. 6 as we describe paradigms. The Cynefin framework is intended to assist managers, policy makers, and others in making important decisions. Developed by Snowden (1999) in IBM it is referred to as a *sense-making device*. Cynefin in Welsh means *habitat*. It is not really a case of two axes/dichotomies. It may be said that the two left-hand methods of decision-making are less ordered than the two right-hand ones. It is also possible to position the four paradigms in this matrix which in this sense serves as a confirmation of the existence of the Emergence Paradigm. As in the four-phase model, Cynefin assumes that four aspects or orientations can be linked in a model. Nevertheless, these models recognize that *switching* to a different orientation does not happen by itself. One can even speak of a fundamental change as a *paradigm shift*. The Cynefin framework sees the transition from a *simple/known/obvious* view of reality to a *chaotic* as extra complex and almost impossible to make (see Fig. 5.8).

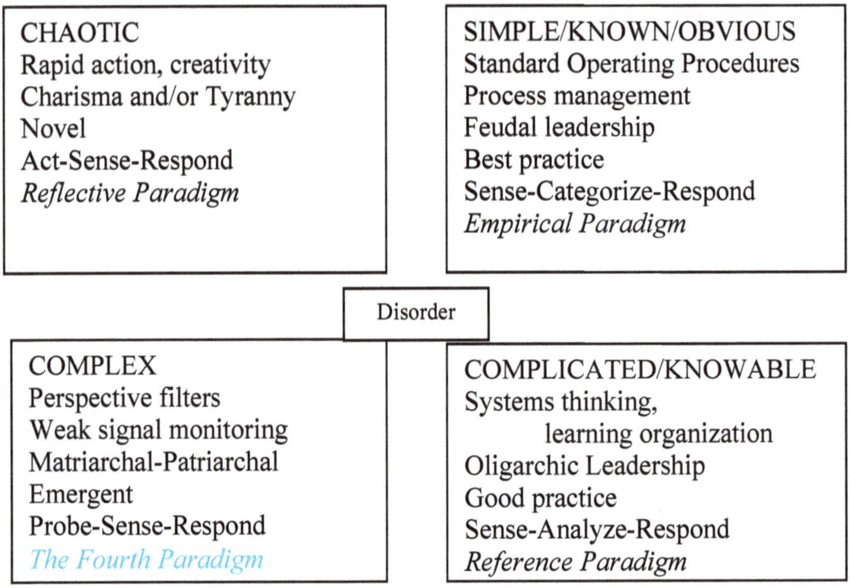


**Fig. 5.8** Cynefin (Snowden 1999)

## 5.6 The Fourth Way of Noordhoek

Noordhoek in his thesis *Trusting Associations* (2019) demonstrates that the three known approaches do show results. However, that they do not function to full satisfaction. The approaches which he describes and refers to as objective, system, and social show major similarities with the above-mentioned Empirical Paradigm, Reference Paradigm, and Reflective Paradigm. In his case descriptions, he demonstrates that over the course of time, by pressure from the surroundings, and the willingness to reach a consensus about what must be understood as quality. There is what he calls a flood of interest in quality. He calls the movement *divergent*. There is room for a fourth approach which he calls *surgent*.

## 5.7 Conclusions

In the above-mentioned paragraphs we have sought for a classification principle for manners of thinking about quality. So, we decided in favor of division into fours. We studied Wilber (2000), Hardjono (1995), *SqEME*®, Bakker and Hardjono (2013), Snowden (1999) and referred to Noordhoek (2019). We always found indications that aside from the three schools of Vinkenburg, a fourth way of thinking about quality was possible.

We come to the following conclusions:

- There is reason to position a fourth way of thinking about quality, in addition to the aforementioned.
- These four ways of thinking about quality relate to each other not as dichotomies, bus as *windows*, and together attempt to describe reality.
- Characteristics of the fourth paradigm could be:
    - Collective values and common language (Wilber)
    - Creativity and risk-taking (Hardjono)
    - Energy exchange and mutual influencing (SqEME)
    - Collaboration (Bakker and Hardjono)
    - Complex, emergent; probe-sense-respond (Snowden)
    - Surgent (Noordhoek)

  We are ready now for a more detailed description of all four paradigms.

## References


Bakker, R. J. M., & Hardjono, T. W. (2013). *Horizontaal organiseren.* Deventer Kluwer, ISBN 978 9013 113662.

Hardjono, T. W. (1995). *Ritmiek en organisatiedynamiek.* The Hague, Kluwer: Vierfasenmodel.

Noordhoek, D. P. (2019). *Trusting associations: A surgent approach to quality improvement in associations*, Ph.D. study, Tilburg University

Snowden, D. (1999). *Liberating knowledge.* CBI Business Guide, London: Caspian Publishing.

van Kemenade, E., & Hardjono, T. W. (2011). Van vier paradigma's naar vier scholen en een Wintercamp, *Synaps, 32*:28–34. Retrieved 4th of March from https://www.vankemenade-act.nl/wp-content/uploads/2017/08/Vier_paradigmas-vier-scholen3.pdf.

van Velzen, R., van Oosten, J., Snijders, Th, & Hardjono, T. W. (2002). *Procesmanagement en de SqEME ® benadering.* Deventer: Kluwer.

Wilber, K. (2000) *A theory of everything: An integral vision on politics, science and spirituality.* Shambala publications.

# Chapter 6
# The Empirical Paradigm, the Reference Paradigm, and the Reflective Paradigm

Introduction

> A paradigm meets the rule of incommensurability, i.e. the paradigms are in principle incompatible with each other, however.....

We have already indicated that the definition of quality as *the quality of an entity* does appeal to us and, above all, we would like to point out that everything may be considered as an entity. We mentioned as an example not only a tangible product, service, process, system, organization, chain, but also society in its totality. For practical reasons, from this chapter on we will mainly focus our discussion on the organization as an entity. The paradigms are not applicable to other entities, so we simply do this to avoid repetition and confusion. We opt for the organization as an entity, since the organization as an entity conceives prerequisites leading to the quality of the available processes. This determines the eventual outcome of the shape of a product, service, or conviction. Organizational quality is also crucial for the chain of processes (the weakest link ...). We also regard quality as the determining factor of the company in its role as a socially responsible organization.

Our conclusion based on the previous chapters is that many attempts were made to define quality and to explain developments in quality science. Only rarely, we found authors who were explicitly aware of the fact that their opinion was colored by the frame of reference on which they based their arguments. Perhaps for this reason, they did not clearly describe this frame of reference, leading to different movements, in which the energy and attention were particularly oriented toward combating one another and not so much toward giving an impulse regarding thinking about quality. This led to several derivatives and applications, such as quality care, quality management, quality systems, quality methods and so on, each with their proponents and opponents. Moreover, we conclude that they nonetheless undertook attempts at making divisions, the importance of which must not be underestimated. We already referred to the classification of Jouslin de Noray or the schools of Vinkenburg. We have analyzed the different views and have defined four paradigms. We



T. Hardjono and E. van Kemenade, *The Emergence Paradigm in Quality Management*, https://doi.org/10.1007/978-3-030-58096-4_6

already mentioned these in the previous chapters. We recognize and acknowledge that: *"A paradigm meets the rule of incommensurability, i.e. in principle that the paradigms are incompatible with each other"*, despite this, we see a correlation and overlap. Hence, we must conclude that the various paradigms are mirror images of each other. On first sight they appear to be the same, yet they turn out to be fundamentally different. In our description of the various paradigms we will address these fundamental differences. For completeness sake, we repeat the names attributed to the different paradigms: the *Empirical Paradigm, Reference Paradigm, Emergence Paradigm,* and the *Reflective Paradigm.* In Chap. 5, we mentioned that division into four is not only a commonly used method and that the number four has a deeper meaning while a fourfold division occurs in many places in nature. Therefore, we believe that the division into four leads to a complete picture. Naming an obvious fifth paradigm does not add anything to our discussion, unless we abandon the division into four.

In this chapter we describe three of the four paradigms. The fourth paradigm will be addressed in the next chapter. For each paradigm we will address the following questions, based on our definition of paradigm: What is quality? What is quality science? We will also discuss our vision of quality science and its representatives as well as the models and tools applied. Finally, we provide an explanation of the paradigm and how it may be recognized in a number of art forms. The addition of the reference of each paradigm to a form of art and music is perhaps debatable. In art it is particularly difficult to recognize movements and to distinguish clear categories. Rather it is, as Rik Spann calls it, *a contextual soup.* It is hard to link a particular art movement to some paradigm, which may vary, based on one's own perspective. Does one examine the way in which art relates to reality or how art is produced? However, via this associative way of thinking we have attempted to make a significant contribution to clarify our thoughts about quality. A clarification with art forms will certainly contribute to this overview. Compare in this respect Pirsig's metaphysics of quality: *"Most empiricists deny the validity of any knowledge gained through imagination, authority, tradition, or purely theoretical reasoning. They regard fields such as art, morality, religion, and metaphysics as unverifiable. The Metaphysics of Quality varies from this by saying that the values of art and morality and even religious mysticism are verifiable and that in the past have been excluded for metaphysical reasons, not empirical reasons. They have been excluded because of the metaphysical assumption that the entire universe is composed of subjects and objects and anything that can't be classified as a subject or an object isn't real. There is no empirical evidence for this assumption at all. It is just an assumption"* (Robert Pirsig, 1991, p. 121).

## 6.1 The Empirical Paradigm

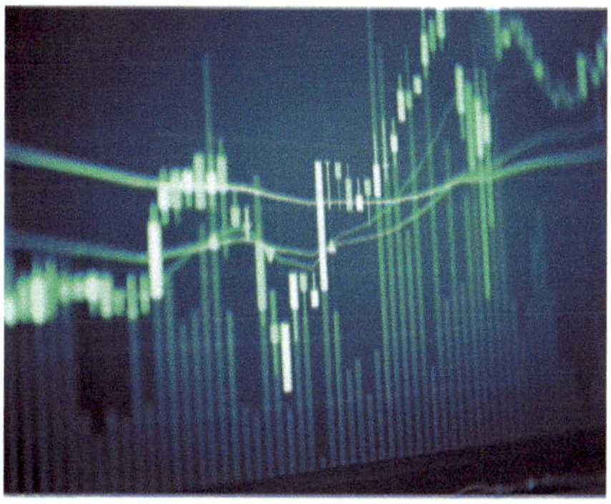

empirical paradigm

We link the Empirical Paradigm to the so-called *positivism*, utilizing the description provided by the Dutch version of Wikipedia:

*"Knowledge may only be acquired by the correct application of the scientific method. Any classical form of metaphysics and other grounds of knowledge are rejected: knowledge is only possible with regard to the world of the phenomena. The term refers to a philosophy based only on observable facts which rejects all metaphysical philosophy and theology, as well as normative knowledge* (which we link to the Reference Paradigm) *or ethics as well as all knowledge that cannot be tested by the senses* (the Reflective Paradigm and Emergence Paradigm). *In positivism, this is linked to the typical belief in the progress of mankind towards a positive, practice-oriented worldview. Positivism is often accompanied by scientism: the belief that science will provide an answer to all problems" (Wikipedia).*

We also opted for the empirical concept since *empirism* as a philosophical movement states that knowledge is based on experience, something often associated with the quality concept. Admittedly, many see empirism as the opposite of rationalism, but here we follow Emanuel Kant, who advocates a synthesis between empirism and rationalism, with a reference to what he refers to as the most fundamental forms of human perception: *space, time, and causality* which determine the way in which we (are able to) think about things in the world. In this context, *rationalism* is the philosophical movement based on the idea that reason is the only or main source of knowledge. According to rationalists, reality contains an inherent reasonable and logical structure that may be read by the understanding without anything else than by thinking itself. Collection of data by means of *observations* and the statistical processing thereof is characteristic for any form of science based on empirical knowledge. Collecting data is a classic component of quality science: measuring is knowing. Since Shewart, statistics has become a fixed part of quality science, as found in the Six Sigma approach today.

*The definition for quality in the Empirical Paradigm is*:

Quality is meeting the specifications.

### 6.1.1 Quality Science

The Empirical Paradigm greatly overlaps with Vinkenburg's empirical school. "The *Empirical school* and the quality science expert work on tangible, specific and current issues related to *product and process* (too many, too late, too long, too expensive, too often). Such issues are of a *technological* nature. The solutions for such issues stand evidence-based, by means of a scientific way in the positivistic tradition". We follow Vinkenburg in this respect. Quality science issues are (relatively) closed, straightforward, and temporary in nature, not socio-psychological in nature and solved in a manner focusing on their technology-related content. Even the Empirical Paradigm focuses on process variation (symptom), the unpredictability of product characteristics, and the uncontrollability of production processes (diagnosis). In principle, issues may be solved by means of variation reduction in process parameters (therapy), by tracing the causes of such variations through statistics and subsequently removing the causes (treatment). The Empirical Paradigm is geared to critically thinking, independent individuals. It is intended for the perfectionists, who want to know all the details; for statistical specialists; for bean counters and accountants. They make use of mathematical know-how. The key elements in this paradigm are (quantitative) data and objectivity. In its most extreme form, the Empirical Paradigm will lead to a circus of measurements and bureaucracy.

### 6.1.2 Instruments

The motto of the Empirical Paradigm is: *Measuring is knowing*. By means of statistics we have the ability to control the production process. The tools entail control charts and statistical process control.

Variation reduction is also one of the principles of Six Sigma.

The view of *measuring is knowing* refers to an objective answer. The judgment of empiric measuring is definitive. Based on hard figures, it is a question of yes or no, right or wrong, sufficient or insufficient. The role of the quality expert is to make these measurements, to perform internal audits, and to report deviations. The old ISO 9001:1987 standard fits this philosophy. It is questionable whether the old ISO standard is a suitable measuring too; furthermore ISO developed since. For this reason, we include the ISO 9000 series under the Reference Paradigm.

### 6.1.3 Representatives

John Locke, philosopher (1632–1704) makes a distinction between aspects that are of primary quality of an object and other aspects that are of secondary quality. Primary quality aspects are things such as hardness, dimensions, shape, motion or rest, and number. These features exist irrespective of whether someone recognizes them or not. The Empirical Paradigm focuses on this primary quality.

Walter Shewart is an important if not the most important founder of quality science (see frame). He not only provided a scientific base to quality science; he also introduced the use of statistics. Deming and Juran should definitely be mentioned as his students. Shewhart (1939) discussed quality management as a process in three steps: specification, production, and inspection. Juran: *"Without a standard there is no logical basis for making a decision or taking action".*

Shewart introduced the *control card* as the true spiritual father of the Plan-Do-Check-Act cycle, even if it is known as the Deming cycle. Shewart's view of science is part of the positivistic tradition and serves as an example for the Empirical Paradigm.

*In his 1931 publication Economic Control of Quality of Manufacturing Product Dr. Walter A. Shewhart (1891-1967) introduced statistical methods to check and improve the quality of goods and services. Thanks to his efforts, quality care received a scientific base. Shewart determined that in industrial processes there are always deviations or variations of the set norm. This happens since there are differences in the quality of raw materials, as well as in the know-how and skills of employees. The application of statistical tools makes it possible to monitor and control the production process. Shewart pictured the trends of the production process on process control charts (control diagrams). He made a distinction between accidental deviations that are within the predetermined margins and the systematic deviations exceeding these margins. As long as the deviations in the production process are within the permissible margins or tolerances there are no problems. Defining the production values in a control diagram reveals the undesired variations and thereby the quality problems. Shewart also asserted that when the production process is under control, future production processes may be predicted within certain limits. The control diagrams have three objectives: defining the norms for the production process, revealing deviations (as well as quality problems) and investigating whether all norms are met. Although Shewart never gained the same notoriety as other gurus, he clearly influenced the work of Wiliam Edwards Deming.* (Emmerik, 2012).

In Wilber's four quadrants the paradigm fits the quadrant of internal/individual, the object is "It". *"It is about behavior that is visible, measurable. It is about science, nature, truth.*" The Cynefin© window related to this assumes the following order: sense–categorize–respond fitting the typification simple/known/obvious, with the warning that in practice the transfer to the chaotic principle is almost impossible to make. In terms of the four-phase model, this "reversal" to an earlier orientation (from an orientation toward efficiency to an orientation toward effectiveness) is possible but often coincides with a loss of capabilities on all fronts. The four-phase model predicts with any "reversal" a loss of capabilities. In other words, the transfer from an orientation toward quality from the Empirical Paradigm to the Reflective Paradigm does not happen without any effort.

The Empirical Paradigm may well be linked to the "correspondence window" of SqEME© and the meaning field "control" as described in *Horizontaal Organiseren* (Bakker & Hardjono, 2013).

### *6.1.4 The Empirical Paradigm and the Four-Phase Model*

As mentioned in the description of the four-phase model (Fig. 5), the four phases may not be regarded as different paradigms. In fact, each phase overlaps with the strategic starting point of the preceding and the following phase. However, the four-phase model says something about the *quality* of the organization and indicates organizational interventions and the related results.

The intervention and the results belonging to an orientation toward efficiency from the four-phase model may be linked to the Empirical Paradigm. In this paradigm, organizational interventions are obvious measures and results, as well as the results of obvious subjects that can be/must be measured.

Seen from the perspective of our own expertise, the four-phase model's orientation toward efficiency is determined by an orientation toward control, combined with an internal focus. A key element is the emphasis on material assets within an organization. The resulting effect depends on our utilization of (material) resources?

Derived from its efficiency orientation, the Empirical Paradigm also has a pitfall. Namely, a rigid orientation toward control results in bureaucracy caused by an inert, guided internal focus.

Depending on the orientation toward control from the four-phase model, one wants to have clarity in the Empirical Paradigm about:

- Which explanatory and prediction models are being used?
- What is the structure and what are the (organizational) relationships?
- Which agreements and dependencies apply?
- Which (measuring error) margins/variations are permitted?

Derived from the internal focus the following must be clear:

- How self-inspection is a part of gathering more knowledge?
- What is the moral/what are the norms and values of the social system within which work and measurements are carried out?
- Which know-how is applied?
- With which resources must the work be carried out?

The above-mentioned organizational interventions/measures must be measurable according to the four-phase model within the Empirical Paradigm

- How clear is the vision and mission of the organization?
- What are the synergy effects?

- How clear is the division of the tasks and responsibilities?
- What is the routine/collected experience?
- What is the profitability of the organization?
- What is the course of the organization?
- What is the reputation of the organization?
- What is the profit and cash-flow generated by the organization?

The above-mentioned measuring points of the four-phase model do all apply to the Empirical Paradigm. The first mentioned measuring point is found in the Reference Paradigm, the last three in the Reflective Paradigm. The middle four items are the core of the Empirical Paradigm, whereby "the profitability of the organization" is the most important measuring point.

### 6.1.5 Organizations

Organizations in this paradigm are hierarchical. The metaphor for such an organization is the machine (Morgan, 1992).

> The mechanical philosophy belonging to the machine metaphor reflects the opinion of most people of what an organization should be. Organization is often considered to be a totality of well-ordered relations between clearly defined parts showing a particular regularity. Thinking is also referred to as *scientific management*. In this context, the faith of 'make plans, organize, check and control' is key.
>
> Kennisbank, Twijnstra Gudde.

The quality function is a staff function, mostly linked under the production manager. Product know-how and know-how of the primary (production) process must be his/her dominant characteristic, know-how of measurement and control technology as well as the application of statistics should be contained. Job titles are the quality department staffed with quality experts. The Dutch Kwaliteits Dienst voor de Industrie (Quality Service for Industry, or KDI), the European Organization for Quality (EOQ), and the American Society for Quality (ASQ) are all examples of organizations where quality experts meet their peers. Characteristic for this organization is an organizational structure along the lines of product groupings such as Food, Agriculture, Automotive, Aerospace. The quality experts in health care have established a separate organization, in which everything points to a dominant Empirical Paradigm (evidence-based machine).

### 6.1.6 The Empirical Paradigm in Religion and the Arts

Atheism fits the Empirical Paradigm. God, Allah, heaven and hell do not exist, since they are not visible or measurable. It is what it is and nothing more.

In conjunction with the discussions about the empirical approach and rationalism *realism* arose, a movement in the nineteenth century visual arts, theatre, music, and literature, attempting to render (social) reality. The "Fisher Girl" (1874) by Ilja Repin is a good example of realistic visual art, just as the major symphonic poems. Many call this qualification "true quality". Realism strives to represent reality as exactly as possible.

Moreover, we see elements of the Empirical Paradigm in modern electronic dance music. Trance is a subspecies of dance emphasizing melody and a euphoric atmosphere. A trace number often begins with a monotonous beat. After a while, more percussion and melody join in. In the middle of the album, a break works toward a climax. Usually, the number of beats per minute varies between 125 and 150. Progressive trance distinguishes itself by a slower rhythm (fewer beats per minute) and a usually long dreamy structure. With hardcore house and speed core the number of beats per minute is quite a bit higher, and may be between the 170 and sometimes even 300. With Extratone the rhythm far exceeds the other forms of music, reaching a number of beats above 1000. The computer makes sure that you stay within these measurable units (when you choose to do so).

**CASE: The Empirical Paradigm in a hospital**

The Dutch Association of Hospitals has issued a memorandum: *Kwaliteit op de Kaart* (NVZ, 2014). It describes how hospitals are required to report on their quality. The following text contains parts of the document.

Health care provided by hospitals in the Netherlands ranks with the top of Europe. Health care quality is high. Access to health care is good. Compared to other Western European countries, health care is affordable. Quality registration and monitoring are quite important for the Netherlands to maintain this top position. Hospitals may use this quality information for improvements, since there is always room for improvement.

Quality information is not only important for hospitals, but also for patients, health care insurers, supervisory institutions and the government. Visibility of quality, therefore, is a spear point of the hospital sector.

Dutch hospitals make quality visible via over 3,400 questions from the program Visible Care and the basic set quality indicators of the Inspectie voor de Gezondheidszorg (Dutch Health Care Inspection, IGZ). In the past year, the number of mandatory indicators about which hospitals must supply data have increased tenfold. Hospitals in particular supply data about the quality aspects structure and process: in 2012 these items together amounted to 96% of the mandatory indicators.

A high score on a structural or process indicator means that the chance of a good result increases. Often the structural and process indicators become less distinctive over the years, since more hospitals manage to get their structure and process in order. The result is what counts and therefore it is important to make the results of health care visible. In recent years, the number of measurements of results in the Visible Care Hospitals and IGZ indicator sets increased from 9 result indicators in 2008 to 140 in 2012.

In the coming years, NVZ strives for a larger part of result indicators that are also reliable and valid.

A summary of our thoughts about the Empirical Paradigm is presented in Table 6.1.

**Table 6.1** The Empirical Paradigm

| | | Empirical |
|---|---|---|
| Questions | What is quality? | Quality is conformance to requirements (Crosby)<br>Quality is compliance to specifications (Gilmore, Levitt) |
| | What is quality science? | Variation reduction in process parameters<br>Quality control<br>Mathematics/Statistics |
| View of quality science | View of measuring | Measuring is knowing |
| | Manner of assessment | Objective based on data |
| | Role of quality expert | Bean counters, accountants, independent people, internal auditors |
| | Representatives | Locke, Shewart |
| Instruments (examples) | Toolbox | SPC, Gauss normal division, Control charts, 6-Sigma |
| | Organizational principles | Hierarchical<br>Machine metaphor |
| Models | Wilber | External, individual, IT |
| | Quinn/Hardjono | Orientation towards control, internal focus |
| | Hardjono (1995) | Emphasis on enlargement of material assets |
| | | ISO9001:1987 |
| | Snowden (1999), Cynefin for decision making | Sense-categorize-respond |
| Art | | Realism<br>Dance music |

## 6.2 The Reference Paradigm

reference paradigm

The word *reference* says that a reference is made in this paradigm. One accepts that quality is indefinable and not as a tangible and exact entity like suggested by the Empirical Paradigm. One accepts that there are constructs created by humans, such as organization, of which one nonetheless intends to measure and possibly improve the quality. Due to lacking specifications, such as the ones on which the Empirical Paradigm bases itself, a norm or model is applied. The (newer) ISO 9000 series, the Malcolm Baldrige National Quality Award Criteria, the European Business Excellence (EFQM) model, and the INK management model are striking examples of the Reference Paradigm. Reference has *referentiality* as its derivative defined as *knowable reality*. A *knowable reality* is defined by means of a frame of reference, that is the general combination of factors forming reality for a person or an entire community. In this case, quality (of an organization) is the measure in which the frame of reference is met.

The Reference Paradigm to a major extent agrees to what Vinkenburg (2006) refers to as the business school, later the normative school. The idea was that quality was determined based on meeting a norm. We partially accept this thought, but we wish to quote the warning voiced by Wikipedia:

*"Normative science" is a science containing or imposing normative statements. The use of the term, due to its contradictory character, is not common in the sciences. Rather, a distinction will be made between "normative research" and "descriptive research", or between the descriptive function of science and the social application thereof. Indeed, science has a theoretical function, but it also fulfills a roll in society. The question is whether this latter role is strictly part of science (forming of hypothesis, analysis, description, critical evaluation and testing etc.) The prescription of norms in and of itself can hardly be regarded as a scientific activity. It fits in with what has been called social constructivism, strongly related to normative science. However for many scientists norms and evaluations cannot be avoided in practice, including for those acting as advisers: political scientists ... It could be*

*argued that this concerns political decisions that are only implemented by scientists. However, most political decisions have come "after" scientific practice, just think of organ transplantation. It can be argued that politicians only take action when medics disagree with each other or when part of the public opinion is protesting. So doctors do make normative decisions. Especially in applied sciences, such as those of physicians, industrialists, agogists, traffic engineers, etc., standards do not arise out of the blue. Whether this legitimizes the term 'normative science', however, remains open to discussion.*

In its most extreme version, the Reference Paradigm leads to rigidity, to model dependence, which as a counter movement may lead to anarchy.

*The definition for quality in the Reference Paradigm is:*

Quality is "fitness for use" or "fitness for purpose"

Norms and models are key elements in the Reference Paradigm. The mentioned models, such as MBNQA and EFQM, are based on an assumption, namely: "the quality of the organization determines the quality of the supplied goods or services". It is an assumption originating from Allied Quality Assurance Publications (AQAPs). The assumption also leads to a new name for the quality position, namely the quality manager. With the *manager* concept, other concepts are introduced, such as *quality care* and *quality system*, with the intermediary form of total quality management (TQM) and integral quality care.

### 6.2.1 Quality Science

If quality management exists, then there can be no quality manager

Quality science strives for total quality management, that ASQ defines as: "a management approach to long-term success through customer satisfaction. In a TQM effort, all members of an organization participate in improving processes, products, services and the culture in which they work".[1] Quality science = quality management. Although it is an attractive idea for managers to assign management tasks to staff services, this is at odds with the concept that it concerns the quality of the organization. It is a responsibility which really should not be delegated. If quality management exists, then there can be no quality manager.

[1] http://asq.org/learn-about-quality/total-quality-management/overview/overview.html.

### 6.2.2 Instruments

Measuring remains important, but in this context measuring means meeting the frame of reference. Its own values are anchored in a model, such as the EFQM model, ISO9001:2015, the Malcolm Baldrige National Quality Award. Even accreditation systems are part of this paradigm, such as that of the Joint Commission International. The assessment model is inter-subjective. In the paradigm of the reference, it is not a matter of an absolute yes or no, as with the Empirical Paradigm, yet the judgment is fixed. Based on knowledge, a third party may form a more widely accepted judgment.

The role of the quality expert is the one who wants to improve the quality in the organization by means of the model

The role of the quality expert consists of one who wants to improve the quality in the organization by means of the model. They work with tools such as self-evaluation, the PDCA cycle (plan-do-check-act), and the Business Balanced Score Card.

One may characterize this as "social constructivism", a movement regarding the learning process as an active process of acquiring knowledge, by which the knowledge originates and is shared with others. One of the underlying principles is that learning is a process of constructing knowledge. In doing so, existing knowledge is further elaborated upon (see wij-leren.nl/sociaal-constructivisme.php).

Social constructivism may also be linked to the term *inter-subjectivity* abhorrent in the eyes of the Empirical Paradigm proponents. The judgment whether something meets the norm or standard, or whether something adheres to the model, is not a (subjective) person's opinion, but the shared opinion of several individuals, preferably contributing various expertise. In the auditing by means of EFQM and INK, it is a standard procedure.

### 6.2.3 Representatives

The philosopher, Sir Thomas More (1478–1535), with his main philosophical publication *Utopia* may be classified in the Reference Paradigm. In this book, More describes an island in the Pacific where everything is organized in the best possible manner, a model society. Juran fits this paradigm, when he focuses more on the customer. Quality is *fitness for use*, in terms of design, conformance, availability, safety, and field use. Yet, he heavily depends on measurements.

Deming is usually classified within this paradigm, since he developed the Plan-Do-Check-Act model, based on Shewart (1939), an iterative model in four steps aimed at realizing continuous improvement in an organization. We will return to this in conjunction with Deming's place in the paradigms.

In Wilber's four quadrants the paradigm fits the quadrant of internal/collective, the object is "we". *"It is about morals, culture, Good".*

The Cynefin© window related to this assumes the following order: probe–sense–respond. The SqEME© window is *Construction* and according to horizontal organization this concerns *construction.*

### 6.2.4 The Reference Paradigm and the Four-Phase Model

Also in this case, the four phases in the four-phase model (Fig. 5) may not be regarded as different paradigms, but comparable to the Empirical Paradigm. The flexibility orientation of the four-phase model delivers organizational interventions and the related outcomes applicable to the Reference Paradigm.

Since each quadrant in the four-phase model has an offshoot forward of a following strategic starting point, and an aftermath of a previous strategic starting point, we will see partial repetitions of interventions and results from other paradigms. In our opinion, this does not affect the *uniqueness* of a paradigm.

For the Reference Paradigm, organizational interventions may be derived from an internal focus, the same witnessed in the Empirical Paradigm, yet not as a measuring point but as a reference point. From this, organizational interventions may be derived from an orientation toward change.

The pitfall related to the Reference Paradigm is *anarchy* caused by an internal focus, affected by inertia and an orientation to change that has led to chaos.

Derived from internal focus witnessed in the Empirical Paradigm, the Reference Paradigm looks for clarity about:

- How self-inspection is a part of gathering more knowledge?
- What is the moral/what are the norms and values of the social system within which work and measurements are carried out?
- Which know-how is applied?
- With which resources must the work be carried out?

The organizational interventions derived from an orientation to change are in the Reference Paradigm.

- Deploying evaluation as a means of enriching knowledge (expanding one's ability to think).
- Deploying (improvement and innovation) projects as a means of reaching a willingness to change.
- Regarding changes (in the market) as opportunities.
- Investigating whether available (tangible and financial) means may be utilized differently.

In the Reference Paradigm, the above-mentioned organizational interventions/measures must contribute to:

- Clarity about the organization's mission and vision.
- Shared (behavioral) values and norms

- Flexibility in the work processes
- Room in the budgets

The four-phase model in the flexibility orientation also mentions "Room for lateral thinking" as a measure point which is more a part of the Emergence Paradigm. Clarity about the division of tasks and responsibilities, the routine/collected experience, and the profitability of the organization, measure points for flexibility following the four-phase model are more part of the Empirical Paradigm than the Reference Paradigm. Maybe with exception for the profitability of the organization in some reference model, for instance, the Malcolm Baldrige Award Criteria and the European Foundation for Quality Management model for business excellence.

### 6.2.5 *Organizations*

Organizations have a clear structure in this paradigm, usually it is a line-staff structure, a tree diagram. According to Morgan (1992), the organizational metaphor is that of the organization as a culture.

> An organization is built on ideas, values, norms, rituals and convictions. The organization as a socially constructed reality, as a pattern of common meaning. By accentuating the symbolic meaning of almost all of life's aspects within the organization, the culture metaphor draws the attention to the human side of the organization. The metaphor refers to another means of achieving organized action: by influencing the language, norms, folklore, ceremonies and other social practices that ensure that the key ideologies, values and opinions controlling the action are disseminated.
>
> (Kennisbank Twynstra Gudde).

The responsibility for the quality of the organization in totality is explicitly based on the leadership of the organization, whereas the responsibility, for example, internal (production and supporting) processes may be delegated. This also applies to the quality of the deliverable products and services, for example of the quality departments. It may be that such a department thinks and operates based on a different paradigm. Usually, this is the Empirical Paradigm.

### 6.2.6 *The Reference Paradigm in Religion and the Arts*

Religious movements such as Islam, Judaism, and Christianity may be regarded as belonging to the Reference Paradigm. The Koran, the Torah, or the Bible constitute the laws which one should obey. For instance, believers have to pray five times a day or go to church on Sundays.

A Greek tragedy is subject to strict guidelines to which it must comply. Each tragedy had the same characteristics: unity of place, time, and action. The prologue related what happened previously. The epilogue explained how the story continued.

There was always a choir singing a song in between the scenes. Each tragedy writer observed these mandatory conventions. In the early Middle Ages, novelists did not strive for originality. Plagiarism was a fairly common phenomenon in this period.

Even classical music is based on many conventions and norms to which composers had to adhere. For example, the classical symphony had four movements; a sonata consists of exhibition, development and reprise; a string quartet consists of two violins, a viola and a violoncello. After the Middle Ages, in the Renaissance, artists enjoyed greater freedom than before. Yet, they did have to conform to global guidelines, such as those for the composition of sonnets or of seventeenth century comedies.

In painting we would like to refer to religious realism as found in "icons". An icon represents a faithful rendering of the prototype, based on a predetermined norm/guideline.

A summary of our thoughts about the Reference Paradigm is presented in Table 6.2.

## 6.3 The Reflective Paradigm

What is your opinion about this?

reflective paradigm

The dictionary defines *reflection* as: observing, reflecting, contemplating. This may be associated with introspection and reflective listening, a counseling technique demonstrating empathy, helping to stimulate the conversation by giving it some depth. Positivism originated from the confrontation between philosophy as well as the success of modern sciences. Philosophers were confronted with empirical sciences appearing to succeed in what they themselves and their (speculative) metaphysical systems were unable to obtain: the acquisition of certain and irrefutable knowledge. This agrees with what the Cynefin model regards as the extra-large barrier between the simple/known/obvious manner of decision-making and the

**Table 6.2** The Reference Paradigm

| | | Reference |
|---|---|---|
| Questions | What is quality? | Quality is fitness for use (Juran)<br>Quality is fitness for purpose (Harvey and Green, De Vries) |
| | What is quality science? | Total Quality Management<br>Quality Management<br>Business Administration |
| View of quality science | View of measuring | What can't be measured is unmanageable (INK) |
| | Manner of assessment | Inter-subjective based on a model |
| | Role of quality expert | Administrators and managers intending to assess the quality of their organization and who put their trust in a model |
| | Representatives | Juran, Deming |
| Models | Wilber | Internal, collective, We |
| | Quinn/Hardjono | Internal focus, orientation toward change |
| | Hardjono (1995) | Emphasis on expanding commercial capability, orientation toward flexibility |
| | | EFQM, ISO9001:2015<br>Malcolm Baldrige Award<br>Accreditations: JCI, NIAZ, HKZ, NVAO |
| | Snowden (1999), Cynefin for decision-making | Probe–sense–respond |
| Instruments (examples)) | Toolbox | Self-evaluation, PDCA, Business Balanced Score Card |
| | Organizational principles | Little forks structure |
| In art | | Conventions in the arts<br>Renaissance: Sonnet<br>Seventeenth-century comedy<br>Religious realism/icons |

chaotic manner. The chaotic manner is offensive to many supporters of the Empirical Paradigm. Consequently, they want to maintain a safe distance from the Reflective Paradigm.

The name *Reflective Paradigm* was derived from the *Reflective School* of Vinkenburg (2006). This paradigm considers quality improvement as a change process and does not wish to adopt a normative point of view. So, it does not tell you how to do it but always asks the question: *"Why is it (not) going well?"*. The essence of the process is: *"Together with others, critically observing one's own actions in order to subsequently make improvements"* (Vinkenburg). Here, the following question applies:

*"What is your opinion about this?"* Ethical and aesthetic observations receive all consideration here. *"It probably meets all of the requirements. But I just don't think it is beautiful"*. Or: *"It does not feel good"*.

Philosophy as a science is a domain of on-going discussion. Admittedly, philosophy of science exists, but that does not make philosophy into a science. Strictly from a normative science perspective, and particularly regarded from the positivism perspective, philosophy does not qualify as a science. Philosophers will certainly have a different idea: It is a lovely example of incommensurabilities. It justifies regarding reflection as a separate paradigm based on the opinion of Frodeman and Briggle. The reasons for this are clear: the philosophy should never have been specialized. "*Rather than being seen as a problem, 'dirty hands' should have been understood as the native condition of philosophic thought – present everywhere, often interstitial, essentially interdisciplinary and transdisciplinary in nature. Philosophy is a mangle. The philosopher's hands were never clean and were never meant to be.*"

As a result of the hyper-sensitivity of opinions in the context and stubborn search for explanatory and predictive models meeting the requirements of positivism, the Reflective Paradigm in its most extreme form can lead to a compartmentalization of thought.

*The definition for quality in the Reflection Paradigm*

Quality is a subjective perception of an event

The definition of quality in the Reflective Paradigm is comparable to Plato's statement: *"Beauty lies in the eyes of the beholder"*. In the words of Pirsig: *"Quality is a subjective perception"*. Values are at the core of this paradigm.

### *6.3.1 Quality Science*

Quality science is no quality control or quality management, but quality *care*. Quality improvement is a process of change. It comes down to connecting the individual truth findings. Vinkenburg, (2006): *"The 'reflective' school has much in common with the 'empirical' school (close observation), but there is one big difference. The 'reflective' school seeks the cause of variations not only in externally perceptible and objectively measurable matters, but also in the human factor, in 'wrong attitudes' and 'unproductive interactions'. This school finds support in the (cultural-philosophical) engagement paradigm, with liveability as an ideal. The role of the quality expert boils down to learn, together with others, to critically examine one's own actions, as a coach, therapist, mediator, friend or court jester"*.

### 6.3.2 Instruments

Measuring is not holy, but *"Who wants to measure something must first know something"*. And: *"Who measures does not get to know anything"*. Whatever is measured happens subjectively. With the Reference Paradigm, the assessment was inter-subjective, and the values were anchored in a model. With the Reflective Paradigm the values themselves are under discussion again. The judgment is only valid for the observer. Eventually, there is a judgment/opinion, but this may change over time.

Instruments are in the field of the above-mentioned connection of truth findings, such as peer audits, inter-vision, time-out, telling *and* listening to stories, a good discussion, second opinion. Pirsig (1991, p. 357): '*What is good, Phædrus, and what is not good—need we ask anyone to tell us these things?*' The conversation is an instrument fitting the Reflective Paradigm. A group discussion evolves about a fundamental, preferably philosophical question. Together, the participants search for an answer to that question.

### 6.3.3 Representatives

With some certainty Plato, Aristotle, Socrates, Pythagoras, Euclides, Ptolemy, Zoroaster, Raphael, Sodoma, and Diogenes may be regarded as representatives.

Pirsig's metaphysics of quality is primarily philosophical and reflective in nature. *"Mental reflection is so much more interesting than TV, it's a shame more people don't switch over to it"*. He also emphasizes the *romantic* aspect of quality: *"Quality is not a thing, but an event"*. The American philosopher Dewey (1859–1952) may (partially) be linked to the Reflective Paradigm, due to his process of *intellection*; the process of reasoning which people need in case their principle to move forward turned out to be unsuccessful, when they want to master the situation once again.

A standard about reflection was written by Schön (1983) with the title *The Reflective Practitioner*. In this book, he distinguishes reflection-in-action from reflection-on-action and reflection on reflection-on-action. The Reflective Paradigm promotes all three of these forms. In Wilber's four quadrants the paradigm fits the quadrant of internal/individual, the object is "I". "It is about art, self, Beauty."

### 6.3.4 The Reflective Paradigm and the Four-Phase Model

We should repeat the comment that the four phases of the four-phase model may not be regarded as different paradigms. The four-phase model, as mentioned previously, does say something about the *quality* of the organization. Derived from the four-phase model, organizational interventions and pertaining results may be attributed to a paradigm.

For the Reflective Paradigm, these interventions and results may be deduced from the orientation on effectiveness from the four-phase model. In this paradigm, organizational interventions must, of course, primarily be understood as a hypothesis to be considered or debated. Particularly the results of this paradigm will be questionable.

The orientation toward efficiency of the four-phase model is determined by an orientation toward control, combined with an external focus (what is the effect upon the surrounding world of whatever is undertaken?).

The central focus is on what the four-phase model refers to as *commercial capability* and what in this context should be understood as the ability to connect with others and to reach agreements.

Moreover, we already pointed out that the orientation toward effectiveness is the mirror image of the orientation toward flexibility (even though they are similar, they are in fact inverted images of each other). In other words: following the four-phase model, the Reflective Paradigm is the mirror image of the Reference Paradigm. This explains why the emphasis is placed on another capability. The focus is on the *commercial* capability rather than the socialization capability at the core of the Reference Paradigm with its social constructivism roots.

The Reference Paradigm has the pitfall of *anarchy*, whereas the Reflective Paradigm has the pitfall of *compartmentalization*, caused by a hyper-sensitivity for external opinions (external focus), as well as the rigidity driving to an extreme with the control orientation.

Derived from the control orientation for the Reflective Paradigm we see the following as topics for reflection:

- Which explanatory and prediction models are being used and why?
- Why is the structure the way it is and why are the (organizational) relationships the way they are?
- What is the basis for the agreements and the applied dependencies?
- Which (measuring error) margins/variations are permitted?

Derived from external focus, in the Reflection Paradigm, one wants to have clarity about:

- What are the anticipated societal developments?
- How will we respond to what happens in the social environment?
- How do we take others along?
- How do we become more concrete?

The four-phase model mentions the following measuring points:

- What are the synergy effects?
- Does the organization possess implementable plans?
- What course does the organization pursue?
- What is the organization's reputation?

- What is the profit and cash-flow generated by the organization?
- Is there room for entrepreneurship (willingness to take risks)?
- Is there market potential?
- Does it lead to product or service-related ideas or concepts?

"What are the synergy effects?" is rather a measure point for the empirical paradigm. "Is there room for entrepreneurship (willingness to take risks), is there market potential, does it lead to product or service-related ideas or concepts?" are rather measure points for the emergence paradigm.

### *6.3.5 Organizations*

Morgan's metaphor of "The organization as a spiritual prison" matches the Reflective Paradigm. The metaphor is based on the view that:

> Organizations are psychic phenomena in the sense that they eventually originate and are supported by conscious and unconscious processes. People may be trapped by images, ideas, thoughts and actions to which these processes give rise. The metaphor shows that although organizations are a socially created reality, people will attribute to such organizations their own existence and ability enabling them in a way to exercise control over their creator (Kennisbank, Twynstra Gudde).

In our opinion, this is a rather negative description which actually only fits the compartmentalized form of the paradigm. If we replace *spiritual prison* by *spiritual environment* and *being trapped* by *being inspired,* then it brings us closer to describing an organization that, in our opinion, thinks and operates based on the Reflective Paradigm.

### *6.3.6 The Reflective Paradigm in Religion and the Arts*

The Reflective Paradigm does not recognize any religious institutions. Religious thinking is individual, subjective, agnostic if you will. The Remonstrants Brotherhood, the Mennonites, and perhaps even the Soefi-movement fit into this category.

In music, this paradigm fits reflections on different variations of the same piece—the Gymnopédies of Satie by Aldo Ciccolini and Reinbert de Leeuw.

Rodin's thinker portrays the practitioner of the Reflective Paradigm in a special way. Raphael's painting Scuola di Atene (1511) depicts ancient Greek philosophy and its practitioners.

Existentialism may be considered in painting, but also in literature. The work of Lucian Freud, for example, is interpreted as existentialism in painting.

**CASE: Reflective Paradigm**

Since the nineties of the previous century, Dutch health care has had a quality instrument called *visitation.* Visitation is *a peer review of the care provided by the relevant professional on location.* Visitation operates according to the principles of peer review. So it is a testing by peers based on equality, confidentiality and expertise. The model is based on learning from each other. So a visitation is not intended to inspect one another. It is not about handing out grades, but it is about holding up a mirror to each other and giving each other tips. In the early years of the instrument's use, this was still voluntary and without strict protocols or standards It was mainly used by professional associations of medical specialists. Initial evaluations were positive, yet too non-committal (insufficiently empirically substantiated, you might say). For example, the Dutch Thorax Surgery Association states: In order to achieve more uniformity within the quality visitations, a number of scientific associations, supported by the Dutch CBO, (quality institute for the health care system), and under the auspices of the Association of Medical Specialists, published the instrument *Draft Handbook Visitation Framework Update* in 2005. In a later stage, visitation was made mandatory by the professional specialist associations. Particularly, Lombarts (2003), has made a case for further tightening up of the system in order to achieve better recommendations and actual quality improvement and prevent non-commitment and hobbyism. In her evaluation of the visitation system, she says:

*Finally, reflecting on the visitation results, the question remains whether they are representative of the actual shortcomings of medical practice or rather reveal the focus of the visitation team and/or visitation program developers. Specialty societies have a great responsibility in selecting the circumstances to evaluate. Assuming that medical specialists will reflect on what they know their peers are going to inspect, the aspects to be evaluated in a visitation process should be selected in the light of their relative importance to the delivery of quality care. (23) Preferably, those aspects are proven to contribute substantially to the enhancement of the quality of patient care (*Lombarts, 2003*).*

Even higher education in the Netherlands has had a visitation system since the nineties. Based on the same peer review principle, visitations were made to colleges and universities at the program level. In the beginning, quality was also strongly dependent on the quality of the committee members. This was one of the reasons why the higher education visitation system was tightened up with a large series of standards that had to be met. In 2002, the old visitation system was replaced by an accreditation system. The *early* visitations in health care and higher education fit in with the Reflective Paradigm; these were later replaced by standards and an accreditation system that fit the Reference Paradigm.

A summary of our thoughts about the Reflective Paradigm is presented in Table 6.3.

**Table 6.3** Reflective Paradigm

| | | Reflective paradigm |
|---|---|---|
| Questions | What is quality? | Quality is not a thing, it is an event (Pirsig)<br>Quality is an event that hits the spot, and that contributes to the quality of life (Vinkenburg) |
| | What is quality science? | Linking of individual truth findings<br>Quality care<br>Change science |
| View of quality science | View of measuring | "Whoever wants to measure something must first know something."<br>"Whoever measures, does not get to know anything" |
| | Manner of assessment | Subjective based on values |
| | Role of quality expert | Anyone who, together with others, is willing to critically examine (learn from) their own actions. And who is willing to fulfill a role as a coach, mediator, therapist, friend or … jester |
| | Representatives | Schön, Pirsig |
| Models | Wilber | Internal focus, individual I |
| | Quinn/Hardjono | External focus, control orientation |
| | Hardjono (1995) | Emphasis on expanding commercial capability; orientation towards flexibility |
| | | "Reflection on reflection-in-action" (Schön) |
| | Snowden (1999), Cynefin for decision making | Sense–analyze–respond |
| Instruments (examples) | Toolbox | Second opinion inter-vision, time-out, stories (telling and listening), a good conversation |
| | Organizational principles | Adhocracy |
| In art | | Rodin, Rafael<br>Existentialism |

## References

Bakker, R. J. M., & Hardjono, T. W. (2013). *Horizontaal organiseren.* Deventer Kluwer, ISBN 978 9013 113662.

Emmerik, R. (2012). *Kwaliteitsmanagement*. Pearson Benelux.

Hardjono, T. W. (1995). *Ritmiek en organisatiedynamiek*. The Hague: Vierfasenmodel, Kluwer.

Lombarts, M. J. M. H. (2003). Visitatie of medical specialists: Studies on its nature, scope and impact. Unniversity of Amsterdam. Retrieved 14th of August 2020 from https://dare.uva.nl/search?identifier=161927aa-aa23-4862-9aa7-df1acfa46ce0.

Morgan. (1992). *Beelden van organisaties*. Scriptum.

Nederlandse Vereniging van Ziekenhuizen, (2014). Kwaliteit op de Kaart. Retrieved 14th of August from https://www.zorgkennis.net/wp-content/uploads/2019/09/ZK-kennisbank-Rapport-Kwaliteit-op-de-kaart-2397.pdf.

Pirsig, R. (1991). *Lila*. New York: Bantam Books.

Schön, D. A. (1983). *The reflective practitioner: How professionals think in action*. London: Temple Smith.

Shewhart, W. A. (1939). *Statistical method from the viewpoint of quality control*, ISBN 0-486-65232-7.

Snowden, D. (1999). *Liberating knowledge*: CBI business guide. London: Caspian Publishing.

Vinkenburg, H. (2006). 'Dienstverlening; paradigma's, deugden en dilemma's'. In *Kwaliteit in Praktijk B1-5*.

# Chapter 7
# The Emergence Paradigm

In the previous chapters, we first gave an introduction to each of the three paradigms, including their definition of quality, and then a description of quality science, its instruments, and representatives. We made a comparison with the four-phase model and gave a metaphor of organization and examples from religion and art. We will do the same with regard to the Emergence Paradigm. The first step is an analysis of the emergence concept, and at the end of the chapter we will discuss the Emergence Paradigm and the path to radical innovation (breakthrough).

## 7.1 Emergence: A Conceptual Analysis

emergence paradigm

Van Kemenade (2019) provides a conceptual analysis of the emergence concept, according to the method of Walker and Avant (2014). This has led to the following features:



T. Hardjono and E. van Kemenade, *The Emergence Paradigm in Quality Management*, https://doi.org/10.1007/978-3-030-58096-4_7

**Table 7.1** Attributes of emergence

| Attributes of emergence |
|---|
| The whole is different than the sum of its parts |
| Interaction/synergy between internal and external elements |
| That occur at the same time (synchronicity) |
| Unpredictable |
| Unexpected |
| Unplanned |
| Leading to a new coherent pattern (novelty) |
| Irreducible to the separate parts |

Emergence occurs in organizations when a 'totality that is new and not merely the sum of its parts' is created. In other words, when something new emerges from a network of interacting internal (entities, parts, agents, individuals, groups) and external elements (the context, environment) in the course of time (novelty). Something new means that the emergence quality or the emergence behavior/pattern is a quality of the organization or the new group that is not present at all or not to such an extent or with that quality in one of the individual elements or in the sum thereof. Nor can it be traced back to the elements from which it originated. The emergent phenomena are present at the same time as the phenomena at the lower level from which they originated (synchronous). The new pattern is coherent, hard to predict, not expected as such and it comes about in an unplanned manner.

In terms of Walker and Avant (2014) we talk about the attributes or characteristics of the concept (see Table 7.1).

This description leads to the following definition:

Emergence is the phenomenon by which a coherent new pattern originates from the network of interacting internal and external elements in the course of time. A pattern that is hard to predict, unexpected and unplanned and that cannot be traced back to the individual elements.

Ablomitz (1939, p. 4) states that "the Theory of Emergence … accounts for the transformation of quantity into quality". Next, as Walker and Avant suggest, Van Kemenade (2019) provides a model case in which emergence occurs.

Model Case: Faculty team of the Integrated Care master's degree program.

The master's degree program Integrated Care at the Utrecht University of Applied Sciences was started in 2007 as an unfunded program with 4 students and 4 part-time professors. The external developments in the sector demand for employees in health care and social wellness, able to make connections, to design innovative interventions, and to realize integrated care. The program which develops such competencies with the participants turned out to be immensely popular. Since the program's establishment, the number of students substantially grew to 133 first-year students and 64 second-year students in the academic year of 2018–2019. Since 2017, the program receives official funding through the Dutch Ministry of Education and Science. In 2018, the program was accredited by the Accreditation Organization of the Netherlands and Flanders (NVAO).

The growth of the program led to major pressure on the faculty team. Via their network, the professors attempted to find more staff for the university of applied sciences (internal

interaction) and beyond (external interaction). Consequently, the number of professors grew exponentially to a team of twenty professors (seven full-time positions).
The team formed in this way differs from the sum of the parts. It has an exceptional diversity in age, experience, and scientific discipline. Yet the organization is marked by harmony, which is new to almost all participants (novelty) and not easy to explain. The harmony exists even if new members are added (synchronicity). There is also some observable synchronicity in the way in which the team members adjust themselves to the curriculum. The new pattern is coherent, which was not to be expected nor predicted (unpredictable). The quality of the collaboration enables continuous learning of the group in its entirety and the participating individuals. It is hardly the result of planned change (unplanned) even though instruments are utilized to maintain collaboration (co-teaching, journal club, and social events).
The result is that the program scores high with the students, even while it has grown so considerably; as evident from among other things the data of the NSE 2018 (score 92 points) and a shared third place of all master's degree programs in the Netherlands rated in the Master's Guide.

The question then is what are the antecedents of emergence (what precedes it) and the consequences (what follows it). The antecedents of emergence are discussed in Chap. 8. There are major differences of opinion regarding what follows emergence. In the literature reviewed (Van Kemenade, 2019) the following possible consequences of emergence are mentioned. Some see a new equilibrium coming into effect. Van Kemenade (2019) sees that—if a new order has come about yet—this is again only (very) temporary. Again and again, the form and nature of the change has to be analyzed. In each instance, the antecedents and consequences of the change must be analyzed. Others see the occurrence of a common identity, radical innovation or breakthrough or the establishment of an ecosystem. (Figure 7.1)

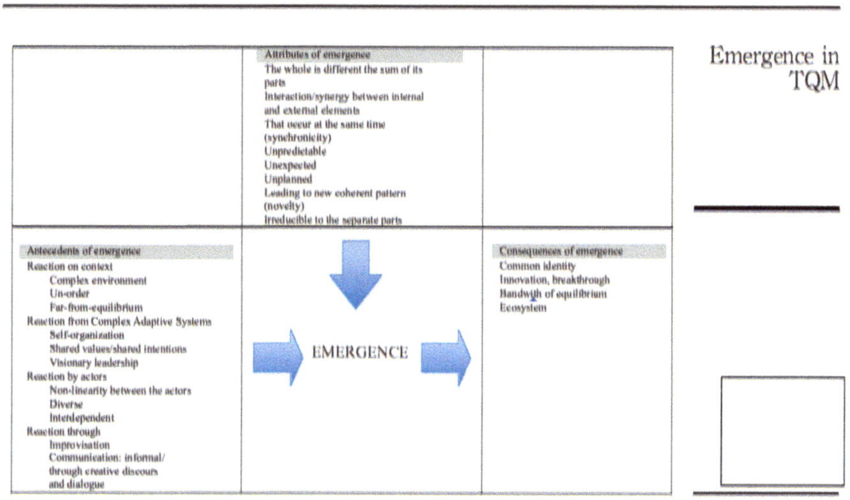


**Fig. 7.1** Antecedents, attributes, and consequences of emergence

Now we can look back at the model case.

Again, the model case: Faculty team of Master Program Integrated Care
When we look at the antecedents and consequences, we see the majority of these in the model case.
In the health care sector it used to be (and there still may be) a matter of un-order and far-from-equilibrium where integrated care is concerned. The program and the formation of a faculty team are a response to the development in that environment. It is even in terms of Snowden & Boone (2007) a complex environment. The development toward the learning community has occurred (and it still occurs) by means of self-organization. Nobody ordered it. Also, it is a case of nonlinearity and diversity of the participants. The development toward a learning community centers around improvisation and creative discourse/dialogue.
Regarding the consequences, it is definitely a case of 'common identity' or 'shared system of meanings', which is evident from the common and individual ambassadorship of integrated care. After the emergence of the learning community we cannot really call this a new order. The environment is in a state of constant movement, just as the (members of) the team. So we should rather talk about a (dynamic) bandwidth within which the result may vary, rather than a (static) new order. This agrees with a number of researchers proposing that complex adaptive systems develop themselves to a critical status between order and chaos (e.g. Packard, 1988; Langton, 1992). "Complex Adaptive Systems 'never get there'. They continue to evolve, and they steadily exhibit new forms of emergent behavior" (Holland, 1992, p. 20).

### *7.1.1 Quality Science*

In the previous chapters we have provided an overview of various approaches to quality. We have presented numerous opinions and ideas, which on the one hand attempt to establish a grip on what is quality *in and of itself*, and on the other hand, how quality may be achieved. In the past, the great variety of opinions and sometimes contradictory ideas formed a reason to several writers to devise some kind of order. Some attempted to base the approach on different objects like product, process, system, chain, and society. The orientation shift was often regarded as a phase in thinking about quality. Others rather tended to think in categories or revolutions. We have clearly presented a case for paradigms. The incommensurability and mutual incompatibility of paradigms immediately explain why the opinions are so diametrically opposed to one another. Also, we have attempted to prove that in addition to the three more or less common paradigms there should be a fourth. This chapter relates to the fourth paradigm, which we have called the *Emergence Paradigm*. Through years of study and authorship, we discovered that the principles of the so-called Emergence Paradigm are very old. We think that even Aristotle made reference to it. After the rise of the Enlightenment and Rationalism, popular attention for this paradigm has been relegated to the background. This may have happened because it could not be reconciled with the scientific view referred to as *positivism*, so strongly based on the views of rationalism and which we have linked to the Empirical Paradigm. The

Emergence Paradigm does not agree with the normative scientific approach of social constructivism. It better matches the paradigm that we call the Reference Paradigm.

Our research revealed that the concept may not be popular in common daily speech, yet that it is used by scientists working on the front lines of their field. Strikingly enough, this also applies within Jazz music and then particularly during jam sessions, with which the concept of improvisation is usually associated. We also found the opinions of Dewey, who by the end of the nineteenth century coined the concept *pragmatic*. The *improvisation* concept and particularly the *pragmatic* concept acquired a sense in daily use, which we would not want to link to a responsible manner of thinking and working on quality. For this reason, we have adopted the concept of *emergence*.

We have noticed that attention to, and views on quality evolved in an emergence way of thinking closely related to *systems thinking* and particularly to *complexity theory*. The thesis of Peter Noordhoek (2019) is based on cases demonstrating an emergent manner of change. It would lead beyond the scope of this book to deal with systems thinking as well as the complexity theory. However, we do assume that the application of the Emergence Paradigm opens the way to disruptive or radical innovation. This is an intriguing fact since quality care has always been at odds with innovation. After all, care for quality has an implicit promise in it, that if applied correctly, there will be no surprises and everything is aimed at measuring, managing, and improving the existing status. At the most, there is only room for incremental innovation. We will refer to this again at the end of the chapter.

As mentioned above, emergence is a key concept to complexity theory. It involves the development of complex, organized systems, exhibiting certain properties that cannot be traced back to the properties of their constituent parts. Due to interaction (new) properties, patterns, regularities, and/or entirely new entities have come into existence.

Emergence is also the process by which new properties come into existence due to the interaction between simple, small entities without such properties, such as the self-organizing ability of ants. Many ants together demonstrate a collective intelligence which individual ants do not have, enabling them to build a substantial termite hill (Quantum Universe.nl).

The Emergence Paradigm is not easily forced into a particular structure. It does not feel comfortable with compartmentalized thinking (in four or five parts). Rather, it is a situation of chaos, to which we must repeatedly give meaning. It relates to taking a decision "based on the best current understanding, having thoroughly inspected everything". See examples in Table 7.2.

The original name of this paradigm (Hardjono & van Kemenade, 2011) was the Pragmatic Paradigm. Indeed, this paradigm is solidly based on the pragmatism of people such as Dewey and Pierce. In 1908, the American professor Arthur Lovejoy pointed out that there were thirteen different meanings of pragmatism. Some of such meanings were even contradicting one another. Lovejoy's statement shows that it is not easy to provide a definition of pragmatism. This is a thought to which all pragmatists will agree. The Emergence Paradigm goes back to the work of John Dewey

**Table 7.2** Examples of emergence

| |
|---|
| In order to gain a better understanding of the emergence concept, we provide yet a number of examples. We see emergence phenomena in physics: |
| *Temperature* |
| Temperature is a measure for the average energy of a great number of molecules. So, we cannot say that a single molecule has a temperature. We are able to measure a temperature when we group many molecules together. In short, temperature is caused by the common interaction of molecules which in and of themselves do not have a temperature |
| *Water liquidity* |
| At a micro level, water consists of $H_2O$ molecules, kept together by hydrogen bridges. Obviously, a single $H_2O$ molecule is not liquid or wet. Yet the interaction (the hydrogen bridges) between the water molecules at a macro level shows all kinds of properties not present at a micro level, such as liquidity and surface tension. Even this example demonstrates that emergence only occurs with a large quantity. Remember that a glass of water contains approximately $10^{24}$ molecules! |
| *Friction* |
| Friction is an emergence force. Elementary particles do not experience any friction. The four fundamental forces between particles retain the (mechanical) energy, while friction causes mechanical energy to be lost and converted into heat. Friction is caused when two surfaces of materials are rubbed together. Since heat is produced due to friction, it is evident that friction in and of itself is not a fundamental force but an emergence phenomenon. |
| *Color* |
| Individual atoms are colorless. The color of an object (consisting of a great number of atoms) depends on the properties of the surface. These properties determine whether the surface absorbs or reflects light of a particular wavelength, due to which a color becomes visible. So color is also an emergence phenomenon |
| There are many other examples of emergence phenomena. Think of a termite hill, consciousness, the internet, a swarm of birds, the crystal structure of snowflakes, the stock market, the traffic at a roundabout, the occurrence of traffic jams and so on. Emergence constitutes a source of inspiration for scientists in many fields of expertise. In recent research, there are even speculations that gravity and space are emergent |

*Source Quantumuniverse.nl*

(1859-1952) as elaborated upon by professor Dr. Sanne Taekema in her inaugural address: *The Problem of Pragmatism*. This philosophy has seven characteristics:

1. It is based on a holistic world view
2. It rejects dichotomies
3. It emphasizes intelligent research
4. It assumes the presence of a temporary status
5. It assumes the presumption of variability
6. It chooses pluralism of methods
7. It chooses the priority of practice

When conducting research, the involvement of both the researcher and stakeholders is regarded as essential, the so-called internal perspective (based on the principle that you cannot intervene).

Typical of pragmatism is that it turns away from the apparently unsolvable problems, to which philosophers have devoted their energy for centuries, while focusing on more everyday problems experienced as such also by non-philosophers. Dewey assumes that (ethical) problems arise in a particular situation related to time and place. In order to understand these problems and to be able to solve them, the situation in which they arise must be studied. We refer to this as the context.

*In order to understand problems and to be able to solve them, the situation in which they arise must be studied (e.g. in participatory research or action research). We refer to this as the context.*

The word *pragmatism* in Dutch has the connotations practical, useful, functional. Yet this is not the sense intended here. As advocated by Rik Spann we chose for the Emergence Paradigm. He referred to an element of the way in which jazz music is created in relation to less improvisation-based musical practices, for example in a classical performance by a symphony orchestra. He also referred to his research at the intersection of organizational studies, musicology, philosophy of art, communication theory and modern natural science, in which the term emergence appears more and more as a key concept.

The Empirical Paradigm may be linked to the so-called *positivism* and knowledge may only be acquired through the correct application of the scientific method in which all classical forms of metaphysics and other grounds of knowledge are rejected, while the Emergence Paradigm precisely constitutes the mirror image. In this context, it is assumed that metaphysical philosophy and theology, as well as normative knowledge (the Reference Paradigm), or ethics and more broadly, all knowledge that cannot be verified by the senses are **not** rejected in advance. Nonetheless, it is assumed that even with the Emergence Paradigm a solid application of scientific methods is required. These may be several (mixed) methods operating side by side. Just as in positivism, the Emergence Paradigm is linked to belief in mankind's on-going progress, yet a progress shying away from radical innovations. The Emergence Paradigm takes the view that properties cannot be explained by the mere reduction of the constituent parts. It is therefore diametrically opposed to the scientism that we mentioned in the Empirical Paradigm. In its most extreme form, the Emergence Paradigm leads to hyper-sensitivity to the opinions of others and to chaos, unpredictable fickleness and hobbyism.

***Definition of quality: Quality is relative and "Quality is never an accident. It is always the result of intelligent effort" (Ruskin, 1891–1900)***

Quality is a relative concept and *"There must be the will to produce a superior thing" (Ruskin).* Quality is not only relative because of its characteristic subjectivity but also because of the sector, its change in time (or not), and its context.

It is possible to mention characteristics of quality, but that still does not capture it. Quality cannot be described in its entirety (Bobbink, 2010). Quality does not *exist*, but it *originates*. More important than attempting to define quality is to determine its (social) function/value. Then, quality is related to virtue. Virtues are the key ingredients of this paradigm.

At first sight, from the perspective of the Emergence Paradigm, quality and quality science appear to be the same as from the perspective of the Empirical Paradigm. Yet, when properly observed, it is exactly the opposite. Where the Reference Paradigm and the Reflective Paradigm are mirror images of each other, the Emergence Paradigm being the mirror image of the Empirical Paradigm. The Empirical Paradigm as a manner of viewing measures the object by means of a predetermined standard. Emergence is viewing in different ways, empirical provides an objective answer, emergence provides several answers. The judgment of empirical measurement is final; by emergence the outcome is tentative. In the case of an empiric judgment, it is a question of yes or no, right or wrong, sufficient or insufficient, based on hard figures. In the case of an empiric judgment, the truth is always in the middle and extremes do not exist. Something is never right or wrong. Non-duality is the starting point here (Buddhism) with at most an ideal far on the horizon (salvation) where many roads lead and progress is the aim. Here, thinking/intellectual capacity is the key element, and money (materialism) is unfortunately a condition that must be given the lowest possible priority.

The way of assessing quality is inter-subjective. Yet related to the context one might call this *inter-objective*. Instruments are the socratic (generative) dialogue and consensus.

Quality science is about creating a quality culture in the organization, about collective consensus using all kinds of existing instruments, including the toolbox from all three other paradigms. Pragmatism is also regarded as *"the art of choosing proven methods that will optimize preferred outcomes"*.

Quality science in this paradigm accepts radical innovation of the object. The role of the quality expert is that of the change agent (van Kemenade, 2011b), open to other visions and models, critical of his own performance and willing to openly dialogue with others. Fisscher describes the role of the quality expert in the foreword to the book Quality Management in Practice of Van de Vaal a.o. (2014) as follows: "Quality experts create room: room to play, communication room and professional room". And: "They are pre-eminently quality experts who, in close connection with strategy development, have an overview of implementation, improvement and innovation, internally and across borders with the outside world in chains and networks, with customer systems and stakeholders. Based on this overview they are able to connect, refer, translate, interview and, traditional though it may sound, provide care".

### 7.1.2 Instruments

Instruments and the emergence paradigm are almost a contradiction. Emergence arises, unplanned, unexpected, hard to predict. This also means that emergence is only *makable* to a limit degree. You may facilitate emergence and perhaps it will happen. Open Space Technology, Appreciative Inquiry and World Café, the Socratic Café as monthly organized by DeGoudseSchool fit as *instruments* in this paradigm. Kemenade (2013, 2014b) developed the ACCRA© model for application in organizations in times of emergent change as an alternative for the PDCA cycle in times of planned change. These are more starting points of action than a tool. In science we recognize Realist Evaluation, Action Research, the participatory research paradigm. The De Goudse School provides conditions, see Sect. 7.7 of this chapter.

### 7.1.3 Representatives

We referred to Dewey, who states that: *"The truth is whatever works"*. And: *"The good is whatever works"*. In Wilber's four quadrants the paradigm fits the quadrant of external/collective, the object is "its". "Its is about context, society". In Hardjono's four-phase model this paradigm fits in the quadrant of external/change. The quadrant which Hardjono has given the name *creativity* fits with an explicit relationship to disruptive innovation, lateral thinking, and investing the intellectual/spiritual capability of an organization. Here applies the Chemistry window of SqEME © just as the window from Cynefin© by Snowden & Boone (2007), and their sequence of action: Act–Sense–Respond. If we follow the book Horizontal Organizing then we end up at *collaboration.*

Scharmer (2000, 2016) explicitly mentions an *emerging future* and applies the U theory. In fact, he provides a roadmap for emergent research, linking this particularly to innovation. (Fig. 7.2)

*Once again Deming*

***We dare to postulate that Deming in particular based his approach on the emergence paradigm***

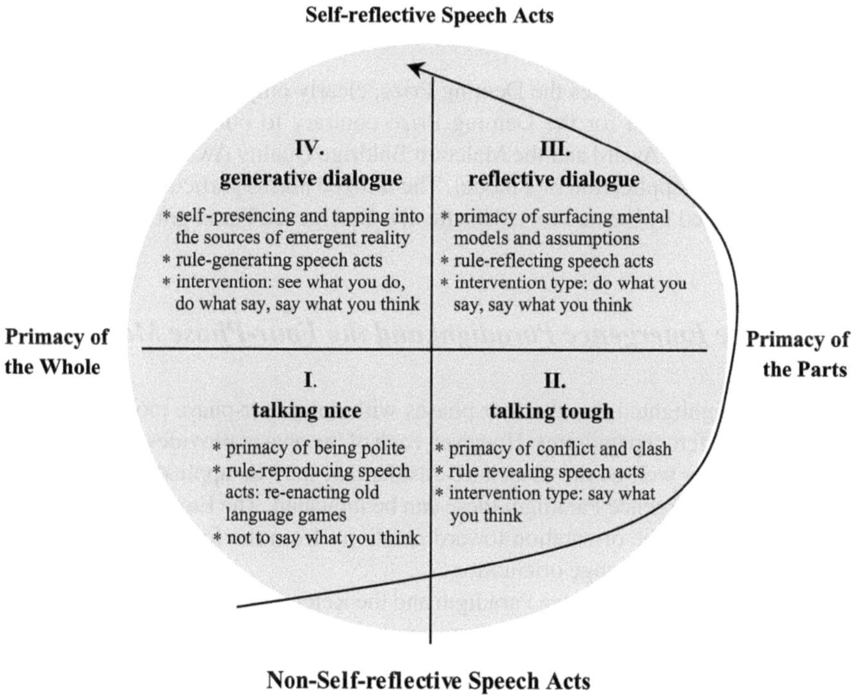


**Fig. 7.2** Otto Scharmer, enacting emerging futures

We mentioned Deming in conjunction with our discussion of the Reference Paradigm due to the PDCA cycle. Deming's name was attached to many publications and we include his fourteen points in Sect. 2.9. It is assumed that Deming first mentioned the PDCA cycle during a lecture in Japan, in 1950. In that version, the individual steps are named *plan, do, check, and act.* In 1993, he modified the cycle in PDSA by replacing *check* by the word *study.* This is because he felt that *check* seemed to refer too much to a form of control, whereas he meant analysis, reflection, and study of processes. Deming also formulated the norms for quality, and in this sense, he is not really a representative of the Reference Paradigm. He did not want to go beyond his fourteen management principles (Sect. 2.9), as formulated in *Out of the Crisis.* When we studied the fourteen points in detail, we came to the conclusion that Deming particularly based himself on the Emergence Paradigm, even though he probably did not know the concept. His ideas became so popular in Japan since they come very close to the Japanese emergence thinking. The latter observation might also hold true for China. May it serve as a warning to the quality world. This statement also explains why Deming and Juran respected one another, but fundamentally disagreed where working on quality was concerned. Now we conclude that their thoughts were based on different paradigms. Juran thought from the Empirical Paradigm and Deming from the Emergence Paradigms. As mentioned above, they are mirror images of one another. One does not realize that a mirror image shows an inverted picture of reality, with everything being the opposite of what is seen and vice versa.

Even when one analyzes the Deming Prize, clearly only to a very limited extent norms are formulated for the Deming Prize contrary to other prizes such as the EFQM Excellence Award and the Malcolm Baldrige Quality Award. The awards are not based on the application of a model. The utilized norms particularly relate to the approach applied by the auditors. This fits the Emergence Paradigm.

### *7.1.4 The Emergence Paradigm and the Four-Phase Model*

Earlier, we highlighted that the four phases within the four-phase model cannot be regarded as different paradigms. However, each of the phases provides organizational interventions, as well as the associated results that may be applied to a paradigm. Even for the Emergence Paradigm these can be indicated. The Emergence Paradigm may be related to the orientation toward creativity from the four-phase model, the result of an external change orientation.

We noted that the Reflective Paradigm and the Reference Paradigm are reflections of one another. The Emergence Paradigm has the Empirical Paradigm as its reflection.

The orientation toward creativity has hobbyism as a pitfall. This may be explained from a hyper-sensitivity external happening and from a change orientation resulting in chaos.

We already observed the organizational intervention derived from an external orientation with the Reflective Paradigm. In the Emergence, these function as stimuli, energy/inspiration sources.

- What are the anticipated societal developments?
- How will we anticipate what happens in the social environment?
- How do we bring others along?
- How do we become more concrete?

The organizational intervention derived from an orientation to change we saw in a suitable form in the Reference Paradigm. For the Emergence Paradigm they may be described as stimuli to:

- Deploying evaluation as a means of enriching knowledge (expanding one's ability to think).
- Deploying (improvement and innovation) projects as a means of reaching a willingness to change
- Which opportunities do these changes (in the market) offer us?
- May available (tangible and financial) means be utilized differently?

As with the other paradigms, this also applies for the Emergence Paradigm, with the above-mentioned interventions and results. The results that one wants to achieve with the Emergence Paradigm are:

- Do all efforts lead to implementable plans?
- Are the participants thinking laterally?
- Is there room for entrepreneurship, willingness to take risks?
- Is there a market potential for expressed ideas?
- Are people able to turn expressed ideas into actual products and services?
- What are the shared (behavioral) values and norms?
- How can we increase flexibility in the work processes?
- How ca we create room in the budgets?

In this context, the four middle concepts belong to the core of the Emergence Paradigm. The aspect of *thinking laterally* belongs to the organization's thinking and intellectual ability. Therefore, it is the most important evaluation aspect. The first-mentioned evaluation aspect "Do all efforts lead to implementable plans?" is also found with the Reflective Paradigm.

### 7.1.5 *Organization*

Just as with the description of the other paradigms, we could limit ourselves to a metaphor such as the one designated by Morgan. In the case of the Emergence Paradigm we see the metaphor "The organization as flux and transformation" as the best fitting. This is based on the idea:

> … that you can't step into the same river twice, since different water is flowing through it all the time. In physics, it is said that in order to unlock the secrets of the universe, we must understand the creative processes that connect the implied as well as the explicit order. If the organizational world is an explicit experience, then we may best understand the organization by unveiling the laws of transformation and change through which this reality unfolds. It is possible to focus your attention on various matters. For example, you might do this by addressing the unconscious as an implicit source of the existence of organizations and asking yourself how unconscious energy is transformed into organizational patterns. You may also want to pay attention to the processes generating the patterns of meaning and symbolic action creating and changing organizations as cultural phenomena. This metaphor heavily relies on insights gained from biology. The organization is regarded as a self-producing system.

*Kennisbank Twynstra Gudde*

Based on the organization as a self-producing system and as a network organization, we think the eight new organizational principles promoted by DeGoudse School is a strong addition to the quoted metaphor with the Emergence Paradigm:

DeGoudse School promotes a mental model[1] that:

(1) Recognizes the mutual dependence (including thinking according to Boerwinkel, 1975)
(2) Strives for intimacy (security and respect); together with the mental model

DeGoudse School bases itself on mutual respect founded on:

(3) Trust (a.o. derived from professional maturity)
(4) Continuous exchange of information (communication about anything); together providing the basis for collaboration.

Makes sure that the organization is designed in such a way that:

(5) It is safe (preventing damage) in combination with
(6) Space (acceptance of variation and variants)

and bases the controls on:

(7) Own responsibility (being transparent)
(8) Accountability (demonstrating that all appropriate measures were taken to avoid waste).

### 7.1.6 The Emergence Paradigm in Religion and the Arts

The Emergence Paradigm has affinity with Buddhist philosophy, but also with LaoTse's writings on Tao, the way. The Emergence Paradigm, unlike the Empirical Paradigm, does not exclude the existence of God.

Magic realism fits in this paradigm: an artistic movement attempting to connect the empirically determinable reality with an "other" or "higher" reality, namely that of a spiritual or psychological order. Not only an attempt is made to evoke

[1] http://www.degoudseschool.nl/organisatieprincipes-21e-eeuw

a particular representation of reality in that higher or psychological order, but also to consciously incorporate metaphysical phenomena in an otherwise closely matching representation. Examples of this are J.L. Borges, Nabokov, Nooteboom (The detour to Santiago). The minimalistic drama productions of the later Samuel Beckett.

In architecture you might think of Gaudi. In painting, the ambiguous work of Dali is a good example or Margritte's oevre with his "C'est ne pas une pipe", or The Lovers Improvisation in music, such as in Jazz, is an example of this paradigm, see e.g. Spann (2018). One might also think of Frank Zappa as a composer, for example of Civilization Phase III:

> The album's storyline was conceived via improvised dialogue involving a series of randomly chosen words, phrases and concepts, which included motors, pigs, ponies, dark water, nationalism, smoke, music, beer and personal isolation. The music was conceived as an opera pantomime, and is dark and ominous. The Ensemble Modern samples allowed the Synclavier to produce richer-sounding music than Zappa's previous works using the machine, which produced the cruder-sounding music on albums such as Jazz from Hell.

Besides for improvisation in his music, Zappa is also known for improvisations in guitar solos, improvisations with the band that he directed, participation of the audience during performances, mutations of his lyrics and improvisational experiments with taped recordings. Improvisation is very popular in modern jazz in general (Zappa & Occhiogrosso, 1989).

Moreover, one might consider experimental theater and theater sport as art expressions of the Emergence Paradigm.

CASE: the Emergence Paradigm.
Since 1951, many organizations in Japan and India have applied for the Deming Prize, established by the Japanese Union of Scientists and Engineers (JUSE). This implies that organizations are able to practice self-evaluation. When organizations apply with JUSE, they must demonstrate that they have determined customer-oriented objectives based on clear leadership, clear management principles, the type and scope of the industry as well as the organizational surroundings.
Moreover, an applicant organization must have implemented TQM in order to achieve the above-mentioned objectives. All of which should lead to excellent results.
The Deming Prize has no elaborate, extremely detailed norms, standards and indicators. An example of this, the element *Status of TQM implementation* may be described as follows: "Elaborate specific TQM activities unique to the applicant organization that have been practiced with an aim to achieve the business strategy. Describe what their features are and how they were promoted. It is effective to divide the description into reasonable chapters and recommends presenting the *Unique and remarkable activity* in an independent chapter" (The Application Guide for The Deming Prize The Deming Grand Prize 2017 For Companies and Organizations Overseas). Subsequently, the following process takes place: "The Deming Prize Examination Committee examines and selects the candidates for the Prize. A document examination will be carried out based on the Description of TQM Practices submitted by the applicant organization. If the applicant organization passes the document examination, an on-site examination will be conducted. The Committee makes judgment according to the evaluation criteria and reports the results to the Deming Prize Committee". Much depends on the discussions of the team performing the on-site examination.

A summary of our thoughts about the Emergence Paradigm is presented in Table 7.3.

**Table 7.3** Emergence Paradigm

| | | |
|---|---|---|
| Questions | What is quality? | Quality is a relative concept; it is related to the user's judgment. Quality is not only relative because of its characteristic subjectivity, but also because of the sector, its change in time (or not) and its context.<br>It is possible to mention characteristics of quality, but that still does not capture it. Quality does not exist, but it originates. More important than attempting to define quality is to determine its (social) function/value. |
| | What is quality science? | Consensus in the collective<br>Quality culture<br>Eclectisch |
| View of quality science | View of measuring | Measuring can be done in many ways<br>Base your approach on variability<br>Pluralistic research methods<br>Primacy of practice<br>No yes/no, provisionality |
| | Manner of assessment | Inter-objective based on virtues |
| | Role of quality expert | Everyone who is open to other visions and models, who exercises self-criticism and is willing to engage in a dialogue with others. Change Agent. |
| | Representatives | Dewey, Deming |
| Models | Wilber | External, collective, its |
| | Quinn/Hardjono | External focus, change orientation |
| | Hardjono (1995) | Emphasis on enhanced thinking/intellectual/spiritual ability, the orientation is towards creativity |
| | | Deming Prize, Socratic dialogue, consensus |
| | Snowden (1999), Cynefin for decision-making | Act–Sense–Respond |
| Instruments (examples) | Toolbox | Open Space Technology, World Café, Appreciative Inquiry<br>ACCRA ©[a]<br>Socratisch Café[b] |
| | Organizational principles | Network organization |
| In art | | J.L. Borges, Nabokov, Nooteboom, F. Zappa, Improvisation in jazz, Beckett |

[a]See e.g. van Kemenade (2013, 2014b)

[b]http://www.degoudseschool.nl/socratisch-cafe-gouda

## 7.2 The Importance of the Context

Introduction

Now that we have described the four paradigms, we return to the question about the relationship between the value orientations and these four. For this purpose, the Emergence Paradigm introduces the *context* concept. For a number of years during our quest, van Kemenade has attempted to treat contextualism as the fourth paradigm (van Kemenade 2014a, 2015b, 2017). He considers the context providing meaning as a[2] quality-science axiom. The axiom of the context determining the meaning supports the choice in ISO9001:2015 to turn the organizational context into a major element of the new standard. This also supports the importance of instruments such as SWOT, PESTLE and *SqEME*. Moreover, attention to context is very crucial in the Emergence Paradigm.

*To understand and be able to solve problems, the situation in which they arise must be studied.*

In our description of the Emergence Paradigm we quoted Dewey, who states that in order to be able to understand and solve (ethical) problems, we need to study the situation in which they arise.

In this final chapter for the epilogue we will discuss the meaning of this situation within the context. In Chapter 2, we stated that the context has no meaning. Yet not all quality-science experts agree. The Western models are still presented as universally applicable, and even worse, as the best models. They are valid here and there, today and tomorrow. Applied worldwide, the standards of the Joint Commission International (JCI) for health care institutions are a good example. However, it is no coincidence that particularly African nations have adopted special variants—irreverently referred to as JCI derivatives—such as Safecare of PharmAccess[3] as well as an individual model of COHSASA[4], better suited to the local requirements. It is also no coincidence that Al-Salmani (2017) in his dissertation indicates that Western quality approaches (ISO and EFQM, treatments from the Reference Paradigm) do not work well in Islamic countries. In particular, the fact that mosque and state are not separate entities imposes different demands, even on quality management.

> *In my work as a consultant and trainer in the field of quality science, I am dealing with a large number of organizations in different countries in the world. I notice that in different contexts different methods (therapies and treatments) are needed to achieve effective quality improvement within organizations. E.g. Alsasmani (2016) shows convincingly that certification schemes and Western approaches to quality do not work in Islamic nations. I have noticed that there are issues depending on the culture surrounding the organization. Sometimes, even the symptoms may differ. Process variation in a Ghanean hospital is not a symptom, but a necessity to be able to meet the demand for health care with extremely*

[2]An axiom is a statement which, though not proven, is accepted as a statement of principle. In and of itself, an axiom serves as a principle of the evidence of other statements.

[3]https://www.pharmaccess.org/update/launch-of-the-safecare-initiative-introducing-standards-to-improve-healthcare-delivery/.

[4]https://www.cohsasa.co.za/

*limited resources. The unpredictability of product properties and the uncontrollability of production processes there is standard and not a problem. In other words, the context provides meaning. Here, the context is defined as the framework within which a specific situation occurs.*
*The importance of the context was previously mentioned in the contingency theory of the fifties. Even then it became clear that Taylor's scientific management did not work, since it ignored the management style and organizational structure influenced by different environmental aspects: the contingency factors Joan Woodward (*1958*) pointed out that there is not one single best way of organizing. Similarly, there is not one single best way of providing leadership (cf. also Hersey* et al, 2001*). Which intervention is most effective at a given moment in the organization depends on the situation in which the organization finds itself. Above all, quality management in an organization should focus on finding a good fit between context and organization. I call this mindset here contextualism. One may think of the internal and external context in this respect. In my contextualism the emphasis is on the external context. (Van Kemenade (2017).*

### 7.2.1 Contextualism

The idea that the context is crucial within quality management is not new. For instance, it was high on the agenda of the contingency theory (see, e.g. Woodward, 1958). The *contingency theory* states that Taylor's scientific management is flawed due to the fact that it does not consider the surrounding factors affecting the organization and its leadership. Woodward (1958) pointed out that there is not one single best way of organizing. Similarly, not one single best way of providing leadership exists (cf. also Hersey et al, 2001 and their situational leadership).

Morgan (2006) states among other things that organizations are open systems forced to adapt to external circumstances. Different environments require different organizations. Differing circumstances force organizations to opt for different approaches to quality. *"No one has the monopoly of the truth"*, as van Kemenade (2015a, b) wrote about the arrogance of some Western quality experts.

*No one has the monopoly of the truth*

Which intervention is most effective at a given moment within the organization depends on the situation in which the organization finds itself? Above all, quality management in an organization should focus on finding a good fit between context and organization. We call this mindset here contextualism. This *mutatis mutandis* also applies to the quality of care provided to an individual patient (see text box).

*Quality management in an organization depends on the context and should above all focus on seeking a good fit between context and organization.*

CASE
**Without context no evidence**
Regarding the illusion of evidence-based practice in healthcare:
The Dutch Council for Public Health and Society writes:
The Hague, June 2017

"*The* proof as a basis for good healthcare is therefore an illusion. Good, patient-oriented care also requires external knowledge that is under-utilized by Evidence Based Practice (EBP): clinical expertise, local knowledge, knowledge provided by patients, knowledge of the context – the living conditions and preferences of patients, the setting in which care takes place—and of the values at stake. Since each decision relates to a specific request for help and since it takes place in a specific context, decision-making in healthcare can be seen as an experiment in connecting different sources of knowledge. Insecurity is inherent to this and must not be denied but should be embraced instead. Every decision may serve and should serve as a learning experience".

Due to the unclear content and shortcomings of EBP the council pleads for *context-based practice* instead of *evidence-based practice*. This is due to the importance of the specific context of the patient and the setting in which various sources of utilized knowledge based on which the decisions are made. This goes further than local implementation of external knowledge. It implies a continuous process of learning and improving together. And it implies another approach to education, research and supervisory practices".

"In the current care system quality monitoring is delegated to third parties and moved away from care professionals. The accent is placed on external responsibility, standardization and control. The Council considers it important to convert this practice into a situation in which healthcare organizations direct good health care while tuning their organization and approach. For this purpose, health care professionals must enter into the dialogue about good healthcare".

It is interesting how contextualism can determine the relationship between the five value orientations and four paradigms.

### *7.2.2 Context and Paradigms*

The strength of a paradigm lies in its context. In a context in which a large fire is raging, or a war is being fought, a value orientation of structure, of order, and of control is most appropriate. In a context in which there is great competition between different parties, a value orientation of success, striving for profit and continuous improvement is appropriate. In a context in which it is internally impossible to achieve success and deliver quality alone, a value orientation of cooperation, community, involvement seems most appropriate. In a context in which organizations have to cooperate externally in order to deliver quality, a value orientation based on attention to context, networking and synergy is appropriate. In a context in which people are looking for meaning, a value orientation of holism, breakthrough, and reaching the essence seems most appropriate.

In a context of relative simplicity and order, we can imagine clear protocols and strict rules offering clarity and solutions (Empirical Paradigm). Where the situation becomes more complex, either the individual or professional will search for solutions beyond the rules (Reflective Paradigm) or less strict models and instructions (Reference Paradigm) from their professional maturity unit. Where there is a lack of order and great complexity, we will not be able to resolve it. There the Emergence

Paradigm is required. This brings us back to Markauskaite & Goodyear's epistemic fluency, the "capacity to understand, switch and combine different kinds of knowledge with different ways of knowing about the world". The context greatly affects when you make which choice (Markauskaite & Goodyear, 2016). In *Twenty First Century Total Quality Management: The Emergence Paradigm* (Van Kemenade & Hardjono, 2019), we referred to this as total quality management.

However, the Emergence Paradigm is the only paradigm placing the context at the center. Vinkenburg (2006) asked why a fourth paradigm is necessary. He wanted to know the problems, symptoms, diagnosis, and therapies within that particular paradigm. The Emergence Paradigm sees a poor relationship with and knowledge of context as a problem for an organization. Symptoms may be resistance from customers to child labor, to competitors who are doing much better, and to no innovations being carried out anymore. The diagnosis is that of an egocentric, narcissistic organization. Therapies allow interactions to take place internally and externally, with collaboration, networking, and radical innovation.

### *7.2.3 Conclusions*

We have discussed the importance of context and the relationship between paradigms and value orientations.

We made following conclusion:

1. A value system determines how the paradigms are treated. From various values this leads to various symptoms, diagnoses, therapies, and treatments.
2. The context gives rise to a preference for a particular paradigm, valid for as long as the context does not change. Inevitably, there is also a certain hierarchy or order of preference to be named. It is the context giving direction and not the paradigm content.
3. Where there is a lack of order and great complexity, we will not be able to resolve this with the three previous paradigms. There, the Emergence Paradigm is required.
4. The art of the quality expert is to make the changes necessary to bring the organization more in line with the requirements of its environment and being able to supervise the application of such requirements, in co-creation with all stakeholders. For this to happen, the quality expert must switch between the paradigms and their instruments. This means that change management is not a foreign object in quality science, but a necessary part of it. The twenty-first century quality manager needs to have change management competencies and attention for context, as stated in the Emergence Paradigm. This is referred to as total quality management.
5. In the Emergence Paradigm the context is central.

In the next chapter we will get deeper into the relation of emergence and radical innovation.

## References


Walker, L. O., & Avant, K. C. (2014). *Strategies for Theory Construction in Nursing.* Harlow: Pearson.

Snowden, D. J., & Boone, M. E. (2007). A leader's framework for decision making. *Harvard Business Review.*

Holland, J. (1992), "Complex Adaptive Systems", *Daedalus*, Vol. 121, No. 1, A New Era in Computation Winter issue, pp. 17–30.

Noordhoek, D. P. (2019). *Trusting associations: A surgent approach to quality improvement in associations*, Ph.D. study, Tilburg University.

Bobbink, H. (2010). 's-Lands Wijs, 's-Lands Eer. *Sigma, 2,* 12–16.

Fisscher, O. A. M. (2014). Forword in: De Vaal, ir. C.D.R.; Pijl, drs.ing. O.J.; van Schijndel. Dr. B.C.; Franciscus, paus, (2015). *Encycliek Laudato Si* (linea 138). Libreria Editrice Vaticana.

van Kemenade, E. A. (2013). De Mythe van de PDCA cyclus, *Sigma, 5:*32–37, Retrieved 4th of March from https://www.researchgate.net/publication/259645824_De_Mythe_van_de_PDCA-cyclus

Van Kemenade E. (2014b). The myth of The PDCA cycle in times of emergent change, *Conference Proceedings EOQ 2014,* Retrieved 4th of March from: http://www.linkedin.com/in/everardvankemenade.

Scharmer, C. O. (2016). *Theory U: Leading from the Future as it Emerges* (2nd ed.). San Francisco, CA: Berrett-Koehler Publishers.

Scharmer, C. O. (2000) Presencing: Learning from the Future As It Emerges On the Tacit Dimension of Leading Revolutionary Change. Presented at *the Conference On Knowledge and Innovation* May 25–26, 2000, Helsinki School of Economics, Finland, and the MIT Sloan School of Management, OSG, October 20th, 2000.

Boerwinkel, F. (1975). *Inclusief Denken.* Bussum: Unieboek b.v.

Hardjono, T. W. (1995). *Ritmiek en organisatiedynamiek.* The Hague, Kluwer: Vierfasenmodel.

Kemenade, E. van, (2014a),Theory C: the near future of quality management, *The TQM Journal*, Vol. 26, Iss 6, pp. 650 – 657, Retrieved 4th of March from https://www.researchgate.net/publication/266967522_Theory_C_the_near_future_of_Quality_Management

Kemenade E. van (2015b) Contextual Divide. An organization's context should determine leadership approach *Quality Progress*/ 2015 November. Retrieved 4th of March from http://www.linkedin.com/in/everardvankemenade

Kemenade E. van (2017), Het belang van de context, *Synaps* 39, pp. 40–44. Retrieved 4th of March from https://www.researchgate.net/publication/317528016_Het_Belang_van_de_Context

Al-Salmani, N. (2017). *Quality management guidelines for Islamic societies.* Amsterdam: VUPress.

Woodward, J. (1958). *Management and technology.* London: Her Majesty's Stationary Office.

Hersey, P., Blanchard, K.H. & Johnson, D. (2001). *Management of organizational behaviour: Leading human resources.* Prentice Hall.

van Kemenade, E. (2015a). De arrogantie van de Westerse kwaliteitszorg *Sigma* 3, juni 2015, pp. 6–10. Retrieved 4th of March from https://www.researchgate.net/publication/315462396_De_Arrogantie_van_de_Westerse_Kwaliteitszorg

van Kemenade, E. (2015b). Contextual divide. An organization's context should determine leadership approach, *Quality Progress.* 2015 November. Retrieved 4th of March from http://www.linkedin.com/in/everardvankemenade.

Markauskaite, L., & Goodyear, P. (2016). *Epistemic fluency and professional education: innovation, knowledgable action and actionable knowledge.* Dordrecht: Springer.

van Kemenade, E. A., & Hardjono, T. W. (2019), Twenty First Century Total Quality Management: the Emergence Paradigm, *TQM Journal*, Vol 31, Issue 2, pp. 150–166. Retrieved 4th of March 2020 from https://www.vankemenade-act.nl/wp-content/uploads/2018/11/TQM-04-2018-0045.pdf

Vinkenburg, H., (2006), 'Dienstverlening; paradigma's, deugden en dilemma's', in: *Kwaliteit in Praktijk B1-5.*

Spann, R. (2018). Improve-isation. Van de groef naar de groove. *Sigma, 2*, 25–29.
Zappa, F., & Occhiogrosso (1989). *The real frank zappa book*. New York: Poseidon Press.

# Chapter 8
# The Emergence Paradigm and the Path to Radical Innovation

## 8.1 Conditions for Radical Innovation: DeGoudse School

In our opinion, an innovation is the application of an invention or discovery leading to a paradigm shift. Sometimes, inventions are *on the shelf* for a long time prior to application and hardly get the attention they deserve. For example, the Nestle group had to take action from above creating room to introduce the Nespresso Machine (Kashani, 2000). Moreover, a clap skate patent has been in existence since 1894. At times, the old system continues for a long time to exist alongside the new invention. It is not uncommon to see an old system undergoing many improvements in competition with innovations on the market. For example, the steamer *Savannah* crossed the Atlantic Ocean for the first time in 1829, but sailboats continued to play a leading role in international shipping until the beginning of the twentieth century. In the course of the nineteenth century sailing vessels were improved and upgraded, as well as the associated infrastructure. We ourselves would not refer to such improvements, also called *incremental innovations*, as innovations. It is hard to maintain that Kodak's films were of a poor quality, just as the mobile phones of Ericsson, Nokia, or the BlackBerry. It is even plausible that much was improved on these devices, until they, like the sailing ships of the nineteenth century, finally became obsolete. Innovations, which we will call radical innovations from now on, in many cases are characterized by the fact that nobody really saw them coming. Also, there was often no clear demand for these innovations, and no one really had an inkling where it would all lead. The car, the Internet, the smart phone, and many breakthroughs—just think of inventions such as sewer construction or penicillin—as examples. Most of these innovations, like Viagra, penicillin occurred unplanned, unexpected. One can only create the circumstances or conditions in which a radical innovation might occur.

Yet from the quality care perspective, these are sometimes painful examples. It is impossible to maintain that the mentioned organizations were of a poor quality. In the Netherlands, Ericsson even played a major role in spreading the INK management model. Nevertheless, we have to conclude that their capacity for innovation proved



T. Hardjono and E. van Kemenade, *The Emergence Paradigm in Quality Management*, https://doi.org/10.1007/978-3-030-58096-4_8

to be insufficient. This conclusion is more painful, precisely since quality care advocates implicitly promise that mistakes will not occur, or at least at a much lower frequently, when a sound process of continuous improvement is initiated and maintained. Continuous improvement of the existing products or services, that is what is at stake here. The Reference Paradigm now tests *existing* norms and models. The Reflective Paradigm poses penetrating questions, but by definition does not provide *any answers* even though explicitly addressing ethics and aesthetics. The Empirical Paradigm measures *what is.* We do not want to diminish the power and value of these paradigms. However, we want to assert that great breakthroughs and groundbreaking ideas should not be expected of them. On the other hand, groundbreaking ideas are to be expected from the Emergence Paradigm. Once these ideas and breakthroughs have come about, they might be better brought to full maturity according to one of the other paradigms. Just as the Reflective Paradigm, the Emergence Paradigm does not provide any answers. Yet it does provide the conditions for ideas to be inspired.

Here, we want to explain what those conditions are in our opinion. The key question is how do you create Morgan's organization, meeting the metaphor "the organization as flux and transformation". We think that the Berenschot seven-forces model for (culture) change offers a stepping stone. These forces are: Vision, capacities, necessity, systems, structures, spirit, and success. (From: *Management van Processen*, Hardjono & Bakker (2001). We add the organizational principles of DeGoudse School, Dewey's characteristics, and the interventions and results from the four-phase model).

The conditions are:

*Care for vision:* Vision provides direction. Using Dewey's characteristics as a test: Take a holistic view of the world, reject dichotomies, put the emphasis on intelligent research, explain precursors, start from the presumption of variability, opt for pluralism of method and for the primacy of practice. Even Deming's fourteen points, which we will not repeat here, may serve as an inspiration for the development.

*Make capacities available.* Capacities make it achievable. Evidently the four-stage model needs material capital, with every item stated on a financial balance sheet. Moreover, the four-stage model refers to commercial capital, which in this context we would like to refer to as "the ability to connect and to achieve transactions". However, it is more important to have access to the so-called *socialization capability.* The ability to work together, to be able to motivate, to let the other person rise above himself. The most important capability in the Emergence Paradigm is thinking and intellectual capability. It is about this capability, locked in the participants' heads within and around the organization, about the ability to mobilize them, stimulating lateral "out of the box" thinking and, of course, learning and showing creativity is vital.

*Clarify the necessity.* Necessity is the power driving the organization's activities. Earlier we mentioned the Emergence Paradigm may be traced back to what the four-phase model refers to as a change orientation in combination with an externally-oriented focus. The two orientations are sources supporting the necessity to move forward. Such as: What are the anticipated social developments? What happens in the social environment? Who are the players? And: Can we be concrete?

Moreover, what do we learn from the evaluation of our actions, how do we utilize projects as a means of increasing the willingness to change, what market developments are anticipated, and how many available resources be used differently?

*Revise the systems.* Systems are the written and unwritten rules through which one makes a career and is (formally and informally) rewarded (and punished).

*Call the structures into question.* Structures anchor paradigms, employment circumstances as well as the organizational requirements. Yet, these stabilizing forces within the organization may be counterproductive to changes. The structures are constantly analyzed in the Emergence Paradigm, while in other paradigms structures are sources of certainty. Introducing the organizational principles of DeGoudse School may be a means to revise the existing systems and structures.

*Show the successes.* Success builds faith in achievability. The success of emergent thinking is measured by the items mentioned at point 8.1.4. It is possible to come up with feasible plans. It is possible to think laterally. There is room for entrepreneurship and a willingness to take risks. It is possible to see market potential and it is possible to show that product, service, or conceptual ideas can actually be realized. However, there are also shared values and norms. Work processes may be flexibly adapted with adequate room in the budgets.

*Stimulating spirit in the organization.* Spirit empowers the organization. The spirit based on the insight that the Emergence Paradigm shows other possibilities than the other paradigms. The Emergence Paradigm may even modify the other paradigms.

Let us look more specific at conditions fostering emergence.

## 8.2 Conditions for Emergence

Van Kemenade (2020) describes the concept of emergence, using Walker and Avant (2014). That description leads to attributes and consequences but we left the antecedents of the concept for discussion here. Antecedents are *"events or incidents that must occur or be in place prior to the occurrence of the concept" (*Walker & Avant, 2014, p.173*).*

*Emergence and Context*

Emergence occurs as a reaction to the context, the environment. Patel and Ghoneim (2011) state: *"Emergence is the apparently sudden and unexpected occurrence of systemic (system-wide) events initiated by the environment that result in a form that is different from the existing form of the system"* (Patel and Ghoneim, 2011, p. 425). Also see Bartezzaghi, 1999 and Weick, K.E., 2000. Sometimes it is stated that emergence is a reaction on crises (Holden, 2005; Chaffin and Gunderson, 2016).

Emergence occurs in particular in a complex environment (El-Hani and Peirrera, 2000; Mitleton-Kelly, 2003; Snowden & Boone, 2007; Ritchey, 2014). El-Hani and Peirrera (2000) state: *"When aggregates of material particles attain an appropriate level of organizational complexity, genuinely novel properties emerge in these complex systems"* (p.133). Roux puts it this way: *"Technological change is conceptualized as the result of a dynamic and complex process emerging from interactions among properties of the constituting components"* (Roux, 2003, p. 11).

Emergence occurs in a context of un-order: *"Un-order is not the lack of order, but a different kind of order, one not often considered but just as legitimate in its own way. Here we deliberately use the prefix "un-" not in its standard sense as "opposite of" but in the less common sense of conveying a paradox, connoting two things that are different but in another sense the same. Bram Stoker used this meaning to great effect in 1897 with the word "undead," which means neither dead nor alive but something similar to both and different from both"* (Kurtz and Snowden, 2003, p. 465).

Emergence occurs in a situation of far-from-equilibrium (Cilliers, 1998; Macintosh and McLean, 2001; MacBeth, 2002, Chan, 2011). *"In 1989, Nicolis and Prigogine showed that when a physical or chemical system is pushed away from equilibrium, it could survive and thrive. If the system remains at equilibrium, it will die. The "far from equilibrium" phenomenon illustrates how systems that are forced to explore their space of possibilities will create different structures and new patterns of relationship"* (Chan, 2011, p.6). Macintosh and McLean state *"A primary concern of complexity theory is with the emergence of order in so-called complex systems which exists far-from-equilibrium in a irreversible medium. Such order manifests itself through emergent self-organisation which occurs as a limited number of simple order-generating rules operate across a densely interconnected network of interacting elements to selectively amplify certain random events through positive feedback"* (Macintosh and McLean, 2001, p. 4). Cilliers (1998) tells us, that *"Complex adaptive systems operate under conditions far from equilibrium, which means there is continual change and response to the constant flow of energy into the system. 'Equilibrium is another word for death'"* (Cilliers, 1998, p.4).

Øgland (2008) talks about the design of a Quality management System and concludes that *"having the organization maintain disequilibrium was a part of the QMS design"* (Øgland, 2008, p. 485).

*Emergence and Complex Adaptive Systems*

The results lead to the question what the relation is between complexity and emergence. Axelrod and Cohen (2000) first define a system, a complex system before taking about complex adaptive systems: "A system includes one or more populations of agents and all of the strategies that those agents employ. A complex system is one in which the actions of agents are tied very closely to the actions of other agents in the system. When the agents in a system are actively trying to improve themselves ("adapt"), then the system is a Complex Adaptive System". Mitleton-Kelly (2003), Snowden & Boone (2007), and Palmberg (2009) consider emergence to be a characteristic of a complex system. The development of emergence here is considered to be the other way around: one of the characteristics of a CAS (Holland, 1992, 1998; Holden 2005; Clayton, 2006; Lollai 2017). Holland (1998) defined CAS as "comprised of interacting agents that constantly and mutually affect each other." Greven (2019) talks about two other important characteristics of complexity thinking: self-organization and emergence. Emergence then is "a situation where out of a certain level of complexity a new pattern of dynamic arises that cannot be described as the sum of the separate parts". However, most scientists consider interacting agents that constantly and mutually affect each to be a prerequisite, an antecedent of emergence. Holden (2005) states in her concept analysis of CAS—also using the method of Walker and Avant—that emergence is the most important consequence of a CAS and that complexity is an antecedent of emergence. Also Holland (1992) speaks of complex adaptive systems (CAS) and mentions a.o. nonlinearity and diversity as characteristics. That supports the idea that complex adaptive systems are antecedent of emergence, they can "produce" emergence (Lollai, 2017). In the words of Clayton (2006), however, "the difficult part, both empirically and conceptually, is ascertaining when and why the complexity is sufficient to produce the new effects" (p. 4).

A similar chicken and egg causality dilemma occurs regarding self-organization. Domingues et al. (2015) study integrated management systems as complex adaptive systems. One of their conclusions is that self-organization inherently arises from management system's integration. Bhaskar (1989) defined emergence as the process where the interactions between actors such as co-workers in an organization, lead to development structures that organize these interactions—self-organization. However, most scientists state that emergence occurs when there is self-organization (Goldstein, 1999; Macintosh and McLean, 2001; Mitleton-Kelly, 2003; Roux, 2003; Hogenson, 2004; Chester and Welsh, 2005; Fabac, 2010; Ellis and Herbert, 2011; Oliveira et al.,2011; Karimi, 2017; Lollai, 2017). Mitleton-Kelly: *"In an organisational context, self-organisation may be described as the spontaneous coming together of a group to perform a task (or for some other purpose); the group decides what to do, how and when to do it; and no one outside the group directs those activities"* (Mitleton-Kelly, 2003, p. 20). The group has shared intentions (Ellis and Herbert, 2011; Taillard et al, 2016). Building shared intentions (Bratman, 2014; Sawyer, 2005) is part of the emergence process. When actors' intended actions are interdependent they develop shared intentions—that is individual intentions that several actors have in common. Palmberg (2009) talks about a shared vision. Van Kemenade and Hardjono (2019) talk about shared values.

*Emergence and the actors*

Emergence presupposes nonlinearity of the relation between the elements (MacBeth, 2002; McCarthy et al., 2006; Wimsatt, 2008). McBeth: *"Non-linear systems produce the important result that small effects can have unforeseen and unforeseeable outcomes"* MacBeth, 2002, p.728).

Emergence occurs by diversity of the elements or actors strengthens emergence. *Emergence is also enhanced by diversity because of the greater interaction and richer patterns* (Holden, 2005). Juriado and Gustafsson (2007) state that the emergent communities of practice in their case study emerged by *"diversity, the number and the fluidity of the individuals and organisations involved"*. Also (2006) mentions diversity.

Emergence occurs if the actors are interdependent (Palmberg, 2009; Ellis and Herbert, 2011; Bratman, 2014).

*Emergence and instruments*

Emergence is fostered by improvisation. Haenisch: "[*T]he term [emergence] points to the characteristics of a collective creativity that cannot be understood in terms of individual accomplishment or ability, but instead as a social practice in which the central authority of artistic authorship is transformed into an aesthetic collaboration, one that is not reducible to a simple 'sum' or chain of individual creative contributions, but that brings about an additional value, thus contributing to an improvisation's success"* (Haenisch, 2011, p.187*)*. Spann (2018) refers to improvisation based on jazz music and argues how important improvisation is for emergence of quality in organizations.

MacBeth (2002) talks about the need for creative discourse as a means for emergence. Snowden and Boone (2007) call it "dissent and formal debate". Leaders must create the conditions: "they have to probe, sense and respond" and "*Because outcomes are unpredictable in a complex context, leaders need to focus on creating an environment from which good things can emerge, rather than trying to bring about predetermined results and possibly missing opportunities that arise unexpectedly"*. Ellis and Herbert (2011) advise management to ensure that lines of communication flow across the network. Fundin et al. (2019) stress the importance of creativity. Van Kemenade and Hardjono (2019) prefer to talk about the creative dialogue. Kelly (1994, p. 469) mentions "growth by chunking"*: "The only way to make a complex system that works is to begin with a simple system that works. Attempts to instantly install highly complex organization—such as intelligence or a market economy—without growing it, inevitably leads to failure."*

Ellis and Herbert (2011) suggest management to apply simple design principles, because they turn into rules. Palmberg (2009) quotes Zimmerman et al. (1998, p. 26) who state: "*It does show that simple rules—minimum specifications—can lead to complex behaviors. These complex behaviors emerge from the interactions among agents, rather than being imposed upon the CAS by an outside agent or explicit, detailed description"*. Table 8.1 summarizes the antecedents of emergence.

This concept analysis discovered that in quality improvement activities we can bring people together and foster interaction. We can exchange *"who we are, what we know and whom we know"* (Lusch & Vargo, 2014, p. 26). We can share intentions and values. We can stimulate creative discourse, dialogue and improvisation. And then—maybe—novelty or even radical innovation emerges.

**Table 8.1** Antecedents of emergence

| Antecedents of emergence |
|---|
| Reaction on context |
| Complex environment |
| Un-order |
| Far-from-equilibrium |
| Reaction from complex adaptive systems |
| Self-organization |
| Shared values /shared intentions |
| Visionary leadership |
| Reaction by actors |
| Non-linearity between the actors |
| Diverse |
| Interdependent |
| Reaction through instruments |
| Improvisation |
| Communication: informal/through creative discourse and dialogue |
| Simple rules |

# References


Axelrod, R., & Cohen, M. D. (2000). *Harnessing Complexity*. New York: Basic Books.

Bartezzaghi, E. (1999). The evolution of production models: Is a new paradigm emerging? *International Journal of Operations & Production Management, 19*(2), 229–250.

Bhaskar, R. (1989). *Reclaiming reality: A critical introduction to contemporary philosophy*. London: Verso.

Bratman, M. (2014). *Shared agency: A planning theory of acting together*. New York: Oxford University Press.

Chaffin, B., & Gunderson, L. (2016), "Emergence, institutionalization and renewal: Rhythms of adaptive governance in complex social-ecological systems", *Journal of Environmental Management,* No. 165, pp. 81–87.

Chesters, G., & Welsh, I. (2005). Complexity and social movement(s): Process and emergence in planetary action systems. *Theory, Culture & Society, 22*(5), 187–211.

Cilliers, P. (1998). *Complexity and Postmodernism. Understanding complex systems*. London: Routledge.

Clayton, P. (2006), Conceptual Foundations of Emergence Theory. In Clayton P. and Davies, P. (eds) *The re-emergence of emergence* (pp 1–34), Oxford; Oxford University Press.

Ellis, B., & Herbert, S. I. (2011). Complex adaptive systems (CAS): an overview of key elements, characteristics and application to management theory. *Informatics in Primary Care, 19,* 33–37.

El-Hani, C. N., & Peirrera (2000). On some theoretical grounds for an organism-centered biology: Property emergence, supervenience, and downward causation. *Theory in Biosciences, 119*(3–4), 234–275.

Fabac, R. (2010). Complexity in organizations and environment-adaptive changes and adaptive decision-making. *Interdisciplinary Description of Complex Systems, 8*(1), 34–48.

Fundin, A., Backström, T & Johansson, P.E. (2019). Exploring the emergent quality management paradigm. *TQM & Business Excellence*. https://doi.org/10.1080/14783363.2019.1591946.

Greven, K. (2019). Theorieovercomplexiteit: het ontrafelen van een vlecht. In Montesano-Montessori, N., Schipper, M., Andriessen, D. & Greven, K. (Eds.), Bewegen in Complexiteit. Voorbeelden voor onderwijs, onderzoek en praktijk, Hogeschool Utrecht, Utrecht.

Hardjono, T. W. & Bakker, R. J. M. (2001). *Management van processen.* Deventer: Kluwer. ISBN 9013034446.

Hogenson, G. (2004), "What are symbols symbols of situated action, mythological bootstrapping and the emergence of the self" *Journal of Analytical Psychology,* Vol. 49, No. 1, pp. 67–81.

Holden, L. M. (2005). Complex Adaptive Systems; a concept analysis. *Journal of Advanced Nursing, 52*(6), 651–657.

Holland, J. (1992). Complex adaptive systems. *Daedalus, 121*(1), 17–30.

Holland, J. (1998). *Emergence: From Chaos to Order.* Reading, Mass: Helix.

Juriado, R., & Gustafsson, N. (2007). Emergent communities of practice in temporary inter-organisational partnerships. *The Learning Organization, 1*(1), 50–61.

Karimi-Aghdam, S. (2017). Zone of proximal development (zpd) as an emergent system: A dynamic systems theory perspective. *Integrative Psychological & Behavioral Science, 51*(1), 76–93.

Kelly, K. (1994). *Out of Control: The New Biology of Machines.* Addison-Wesley, New York: Social Systems and the Economic World.

Kurtz, C. F., & Snowden, D. J. (2003). The new dynamics of strategy, Sensemaking in a complex and complicated world. *IBM Systems Journal, 24*(3), 462–483.

Lollai, S. A. (2017). Quality Systems. *A Thermodynamics-Related Interpretive Model, Entropy, 19*(418), 22–42.

Lusch, R., & Vargo, S. (2014). *Service-dominant logic: Premises, perspectives, possibilities.* Cambridge: Cambridge University Press.

MacBeth, D. (2002). Emergent strategy in managing cooperative supply chain change. *International Journal of Operations & Production Management, 22*(7), 728–740.

McCarthy, I. P., Tsinopoulos, C., Allen, P., & Rose-Anderssen, C. (2006). New product development as a complex adaptive system of decisions. *Journal of Product Innovation Management, 23*(5), 437–456.

Mitleton-Kelly, E. (2003), *"Ten principles of Complexity and Enabling Infrastructures",* available at https://www.researchgate.net/publication/38959109 (accessed 31st of January 2019).

Øgland, P. (2008). Designing quality management systems as complex adaptive system. *Systemist, 30*(3), 468–491.

Oliveira, O. L., Rezende, D. C., & Carvalho, C. C. (2011). Horizontal inter-organizational networks as a complex coevolutionary adaptive system: the case study of a supermarket network. *Revista de Administração Contemporânea, 15*(1), 67–83.

Palmberg, K. (2009). Complex adaptive systems as metaphors for organizational management. *The Learing Organization, 16*(6), 483–498.

Patel, N., & Ghoneim, A. (2011). Managing emergent knowledge through deferred action design principles. *Journal of Enterprise Information Management, 24*(5), 424–439.

Ritchey, T. (2014). On a morphology of theories of emergence. *Acta Morphologica Generalis: Swedish Morphological Society.*

Roux, V. (2003). A dynamic systems framework for studying technological change: Application to the emergence of the potter's wheel in the southern levant. *Journal of Archaeological Method & Theory, 10*(1).

Snowden, D. J., & Boone, M. E. (2007). A leader's framework for decision making. *Harvard Business Review.*

Spann, R. (2018). Improveisation. Van de groef naar de groove. *Sigma* (2), 25–29.

van Kemenade, E. A. (2020). Emergence in TQM: A concept analysis. *TQM Journal, 32*(1), 143–161. Retreived 4th of March 2020 from https://www.vankemenade-act.nl/wp-content/uploads/2019/10/DEF-10-1108_TQM-04-2019-0100.pdf.

Kemenade E.A. van & Hardjono, T.W. (2019), Twenty First Century Total Quality Management: the Emergence Paradigm, *TQM Journal*, Vol 31, Issue 2, pp. 150–166. Retrieved 4th of March 2020 from https://www.vankemenade-act.nl/wp-content/uploads/2018/11/TQM-04-2018-0045.pdf.

Walker, L. O., & Avant, K. C. (2014). *Strategies for Theory Construction in Nursing.* Harlow: Pearson.

Wimsatt, W. C. (2008), Aggretativity, Reductive Heuristics for Finding Emergence, in Bedau, M and.

Zimmerman, B., Lindberg, C., & Plsek, P. (1998). *Edgeware: insights from complexity science for health.*

# Chapter 9
# Epilogue

We introduced this book by asking questions: Why is it that after some initial enthusiasm for quality management, the attention of the top of organizations wanes again? Why do we see many references to Robert Pirsig's *Zen and the Art of Motorcycle Maintenance*, yet so little actual application? Why does it seem that attention for quality and innovation are total opposites? Why is attention for quality particularly hard to achieve within the health care sector? And why does health care stick to approaches different from those applied in other sectors? It appears as if the available set of tools is insufficient. Of course, existing instruments may need improvement, but does quality science itself have a potential for innovation? We went looking for those answers. It is now clear that not knowing the Emergence Paradigm is the answer to all these questions.

We observed that quality is the quality of an entity. The way in which quality is established, improved, and safeguarded determines the meaning of quality science, quality assurance, quality system, quality management, and so on. We have not been able to identify the essence of quality. In everyday speech, quality is also associated with excellence, perfection, beyond all expectations, the totality of positive characteristics. Probably, quality stands above all for the value that stakeholders attach to an entity.

We think we can deduce from different ideas, theories, ways of looking and models that four paradigms may be distinguished. These four consist of the Empirical Paradigm, the Reference Paradigm, the Reflective Paradigm, and the Emergence Paradigm. We opted for the paradigm concept since the incommensurability or mutual incompatibility, not only explains the contradictions in the different approaches, but it also explains why the different conceptions clash. We have gone so far as to associate each paradigm with a different form of scientific practice. By using the four-stage model, we have shown that aspects in the form of organizational interventions and the designation of performance indicators (in the description we called them results) sometimes overlap with other paradigms. Yet because the paradigms


T. Hardjono and E. van Kemenade, *The Emergence Paradigm in Quality Management*, https://doi.org/10.1007/978-3-030-58096-4_9

are mainly mirror images of each other, interventions and performance indicators may have different meanings.

We saw that long before the origins of recognition by guilds, product and process orientation were at the center and thereby only the Empirical Paradigm had some validity. In those early days, when business transactions took place directly between the individual producer (the blacksmith, the tailor, the farmer, etc.) on the one hand and the individual user on the other, the users' needs were easily ascertainable. (Yet it is questionable whether what they wanted was "quality".) Over time, more suppliers became involved in the supply chain from producer to consumer. So, contacts between producer and consumer became less direct. When the guilds arose, they actually had a quality-controlling function for a particular profession. Guilds established the quality norms for production, for use of materials, as well as for the end products. The key concept in this quality control approach was the craftsmanship of the maker(s). Product quality was checked by institutions established particularly for this purpose. Shewhart (1931) was the first modern expert wrestling with the concept *quality*. He distinguished two aspects: the objective aspect (an objective reality independent of the existence of men) and a subjective aspect (what we think, feel, or sense as a result of the objective reality). Shewart opened the way to other paradigms. In the United States and in Europe, the focus was placed on the quality of the organization. This resulted in the introduction of the AQAP, ISO 9000, the Malcolm Baldrige Award Criteria, and the European Business Excellence (EFQM) model, the Reference Paradigm. Unnoticed by the West, Japanese thinking was based on the Emergence Paradigm. Introducing Pirsig, we became familiar with the Reflective Paradigm.

Since then, we have had many different views, as can be seen in the definition of quality in ISO 9000:2005: *"Quality is the set of characteristics and characteristics of a product or service that is important for meeting defined and self-evident needs"*, or *"conformance to requirements"* (Crosby 1979)" or *"conformance to specifications"* (based on e.g. Levitt, 1972). Often, instead of specification or requirement the terms *norm* or *standard* are utilized. The definition of standard according to ISO is: *"document, established by consensus and approved by a recognized body, that provides, for common and repeated use, rules, guidelines or characteristics for activities or their results, aimed at the achievement of the optimum degree of order in a given context" (ISO/IEC Guide 2: 2004).* The clearer and the more measurable the norms, the better we get a grip on the question whether an object meets the norms determining quality. Product requirements may be translated into process requirements, such as requirements of the input and output of the process steps. It concerns: *making it right the first time* (Crosby, 1979). Then it is said: "Once we have control of the process by which the products are made, we no longer have to check every product".

Others put the emphasis on the demands imposed by the user. For example, Garvin (1984) distinguishes the user-oriented approach. Oakland (1993) speaks about "meeting customer requirements". Also within this approach fits Juran (1980), who defines quality as: "fitness for use". See Juran, Gryna, and Bingham (1974) and Juran and Gryna (1993). Feigenbaum (1999), states that: *"the key is transforming quality from the past emphasis on things gone wrong for the customer to emphasis upon the increase in things gone right for the customer, with the constant improvement in the sale and revenue growth"* (p.376). In this context, John Guaspari gives a paraphrase to the transcendental approach of quality: *Quality? I know it when I see it.* Which he changes into: *Quality: You will know it when* ***they*** *see it.* The highest quality has the product or service that best meets the expectations of a particular customer (group of customers). Deming observes that quality goes even further than that: *"Quality should be aimed at the needs of the customer, present and future".*

We preferred the term *stakeholders* and wondered to which stakeholder(s) we should listen if we want to determine quality. In the INK management model, we see that a balance must be struck between *customer appreciation, employee appreciation, community appreciation,* and *management appreciation. "As an organization is able to do more, the value of the organization grows. The value will be different for each interested party (shareholders, management, employees, customers and other stakeholders) and will also be expressed in other dimensions (market capitalization, sales value, status value, value as living- and working environment, value as supplier, value as customer, news value* etc. etc.*)"* (Hardjono 1995). So we see that there are various coexisting stakeholders (customers, personnel, management, shareholders but also suppliers, partners) each with their own norms.

Some writers feel that it is impossible to determine quality in view of the many stakeholders. Pirsig (1974, p. 215) literally says: Quality *is not a thing; it is an event* and Vinkenburg (2007) refers to the implication of this based on events in the relationships and interaction between people. Pirsig makes a plea for allowing emotions in this context. When he discusses fixing his motorcycle, he writes that the mechanics were just like spectators and there was no identification with the work. What the mechanics were lacking, according to Pirsig, was what he himself had experienced when he bent over his motorcycle, namely *feeling*. This entails enthusiasm, empathy, and passion to do a good job, as well as patience or taking the time to do it. Such passion more and more seems to be disappearing from society, he observes. According to Pirsig, most people operate mainly based on their mind. He notices this while fixing his motorcycle. He gets so absorbed in his work that he does not experience the engine as a separate object but feels as if he is merging with it. He loses himself in the job. We know the phenomenon worked out in greater detail by Csikszentmihalyi as Flow. Motorcycle maintenance in the form of zen. It's all about finding the right balance: Allowing the feeling (again) and combining it with the ratio. It is a question of daring to look at reality from several perspectives.

In any case, quality remains a relative concept. Quality is not only relative because of its characteristic subjectivity but also because of the sector, its change in time (or not), and its context. It is possible to mention characteristics of quality, without actually capturing it. It can never be described in its entirety. Yet, more important than attempting to define quality is to determine its (social) function/value.

Inspired by Pirsig, Van Kemenade et al. (2008) argue for the introduction of values in the discussion about the quality definition. In order to know what a person considers quality; it is important to know the values held by the person. The same thing applies to an organization or a system. Where behavior and body language are the manners in which a person acquires values, it is the organizational culture with its agreements, rituals, and procedures that teaches an employee the organizational values. This thought also fits the conclusion of Weggeman (2008) that professionals in organizations cannot be controlled by rules and procedures, but that it is important to create *shared values*. Pirsig (1991) in his second book Lila comments: *"Quality. Quality was value. They were the same thing" (o.c., p. 66/67).*

In our quest for the four paradigms, we wanted to explicitly include this view, just as another Pirsig derivative: his hierarchy of values—inorganic values, organic values, social values, and intellectual values. Hardjono (1995) translates this into four capabilities, which every individual and every organization should have: material assets, commercial capability, socialization capability, and intellectual capability. He has worked out the details in his four-phase model stating that the entirety is more than the constituent parts.

Although we cannot define quality for everyone in the same way, we are able to adjust ourselves to it. This convinced us that there must be several paradigms based on which we should view quality. A collection of paradigms providing room to what is rational and what is emotional and what could be a topic for dialogue. It's about constantly rediscovering what's valuable, using all the models and tools at your disposal depending on the organizational context. Our view opens the way to the Emergence Paradigm and a path to radical innovation.

*In summary: We speak of paradigms and not of schools. We speak of four paradigms, not as the extremes of two dichotomies, but as windows to study reality. The paradigms are not different regarding the object studied, but in the way they are studied. We call them the Empirical Paradigm, the Reference Paradigm, the Reflective Paradigm and the Emergence Paradigm, whereby the latter also paves the way to radical innovation (see Table* 9.1*).*

**Table 9.1** The four paradigms in thinking about quality

| Appendix 1 | | Empirical | Reference | Reflection | Emergence |
|---|---|---|---|---|---|
| Questions | What is quality? | Quality is conformance to requirements (Crosby)<br>Quality is compliance to specifications (Gilmore, Levitt) | Quality is fitness for use (Juran)<br>Quality is fitness for purpose (Harvey and Green, De Vries) | Quality is not a thing, it is an event (Pirsig)<br>Quality is an event that hits the spot, and that contributes to the quality of life. (Vinkenburg) | Quality is a relative concept, it is related to the user's judgment. Quality is not only relative because of its characteristic subjectivity, but also because of the sector, its change in time (or not) and its context<br>It is possible to mention characteristics of quality, without actually capturing it<br>More important than attempting to define quality is to determine its (social) function/value |
| | What is quality science? | Variation reduction in process parameters, Quality control, Mathematics /Statistics | Total Quality Management<br>Quality Management<br>Business Administration | Linking of individual truth findings<br>Quality care<br>Change science | Consensus in the collective<br>Quality culture<br>Eclectic<br>Radical Innovation |

(continued)

**Table 9.1** (continued)

| Appendix 1 | | Empirical | Reference | Reflection | Emergence |
|---|---|---|---|---|---|
| View of quality science | View of measuring | Measuring is knowing | What can't be measured is unmanageable (INK) | "Whoever wants to measure something must first know something." (Hoebeke). And: "Whoever measures, does not get to know anything" (Noordvliet). | Measuring can be done in many ways<br>Base your approach on variability<br>Pluralistic research methods<br>Primacy of practice<br>No yes/no, provisionality<br>Participative |
| | Manner of assessment | Objective | Inter-subjective | Subjective | Inter-objective |
| | Role of quality expert | Bean counters, accountants, independent people, internal auditors | Administrators and managers intending to assess the organization "from the top down" and who put their trust in a generic, abstract and timeless model | Anyone who, together with others, is willing to critically examine (learn from) their own actions. And who is willing to fulfill a role as a coach, mediator, therapist, friend or … jester | Everyone who is open to other visions and models, who exercises self-criticism and is willing to engage in a dialogue with others<br>Change Agent |
| | Representatives | Shewart, Juran, Pirsig | Juran, Deming | Dewey, Pirsig, Rodin, Rafael | Dewey, Deming |
| Models | Wilber | External, individual, it | Internal, collective, We | Internal, individual, I | External, collective, its |
| | Quinn/Hardjono | Orientation towards control, internal focus | Internal focus, change orientation | External focus, control orientation | External focus, change orientation |
| | Hardjono (1995) | Emphasis on enlargement of material assets | Emphasis on enlargement of socialization assets | Emphasis on enlargement of "commercial" capability; the ability to create connections | Emphasis on enhanced thinking /intellectual ability |

(continued)

**Table 9.1** (continued)

| Appendix 1 | | Empirical | Reference | Reflection | Emergence |
|---|---|---|---|---|---|
| | | ISO9001:1984 | EFQM, ISO9001:2015, Malcolm Baldrige Award Accreditations JCI, NIAZ, HKZ, NVAO | "Reflection on reflection-in-action" (Schön) | Deming Prize, Socratic dialogue |
| | Snowden (1999), Cynefin for decision making | Sense-categorize-respond | Act-sense-respond | Sense-analyze-respond | Probe-sense-respond |
| Instruments (examples) | Toolbox | SPC, Gauss normal distribution, 6 Sigma | Self-evaluation, PDCA, Business Balanced Score Card | Second opinion<br>Inter-vision, peer audits<br>Time-out<br>Stories (telling and listening), A good conversation | Holistic, intelligent study<br>Fractals<br>Consensus<br>Rejection of dichotomes<br>Appreciative Inquiry<br>ACCRA ©<br>Socratic Café<br>Open Space Technology<br>World Café<br>Action research |
| | Organizational principles | Hierarchical | Bureaucratic/little rakes | Adhocracy | Network organization |
| In art | | Medieval novel<br>Traditional classical music<br>Greek tragedy | Renaissance: Sonnet<br>Dance/trance music<br>Seventeenth-century comedy | Existentialism<br>Examples of Rodin, Rafael Cicoolini and De Leeuw about Satie | J.L. Borges, Nabokov, Nooteboom,<br>Improvisations in Jazz<br>Frank Zappa<br>Beckett |

## References

Crosby, P. B. (1979). *Quality is free.* New York: McGraw-Hill.

Feigenbaum, A. V. (1999). The new quality for the twenty-first century. *TQM Magazine, 11*(6), 376–383.

Garvin, D. A. (1984). What does 'product quality' really mean? *Sloan Management Review*, 25–43.

Hardjono, T. W. (1995). *Ritmiek en organisatiedynamiek.* The Hague: Vierfasenmodel, Kluwer.

Juran, J. M., Gryna, F. M., Jr., & Bingham, R. S. (Eds.). (1974). *Quality control handbook* (3rd ed.). New York: McGraw-Hill.

Juran, J. M. (1980). *Upper management and quality.* New York.

Juran, J. M., & Gryna, F. M. (1993). *Quality planning and analysis* (3rd ed.). New York: McGraw-Hill Book Company.

Levitt, T. (1972). *Production-Line approach to service.* Harvard Business Review.

Oakland, J. S. (1993). *Total quality management, the route to improving performance.* Oxford: Butterworth-Heinemann Ltd.

Pirsig, R. M. (1974). *Zen and the art of motorcycle maintenance.* New York: Bantam Books.

Pirsig, R. M. (1991). *Lila: An Inquiry into morals.* London: Transworld Publishers Ltd. ISBN 0-593-02507-5.

Shewhart, W. (1931). *Economic control of manufactured product.* New York: D. Van Nosttrad Co., Inc. Reprinted in Sower, V.; Motwani, V. And Savoie, M. (1995). *Classic readings in operations management* (pp. 191–214). Ft. Worth, Texas: The Dryden Press.

Snowden, D. (1999). *Liberating knowledge*, CBI Business Guide, London: Caspian Publishing.

Vinkenburg, H. (2007). Open brief van Huub Vinkenburg op de vorige Synaps. In *Synaps,* 22.

Weggeman, M. (2008). *Managing Professionals? Don't! How to step back to go forward.* Warden Press.

# Chapter 10
# Colleagues' Responses

We submitted our draft manuscript to eight quality science experts. As stated in the introduction, we are looking for a dialogue and are not pretending to have all the answers. Therefore, we are very pleased that they all responded by providing valuable ideas to supplement our manuscript. We have included their responses in their entirety in our book as additional material for a dialogue on emergence.

## 10.1 Quality Lexicon (Huub Vinkenburg[1])

In quality science (such as science and technology), I am interested in improving the characteristics of objects. This doesn't happen from a curiosity about the facts, but from a sense of concern. In this respect, I think, Hardjono and I differ from one another. That's why I prefer to speak not of quality management, but of quality assurance. I mainly limit my contribution here to the *language use*. See the following also as an impetus to a *quality lexicon*.

- *Quality*. I consider the term quality (lesson from Analytical Philosophy) as an *indefinite concept*. It cannot be defined; the meaning differs per context. Any attempt to define the concept of quality will fail, even here.
- *Quality Science*. The term *quality science* (as a discipline, and therefore *other than quality management)* refers to the accumulated knowledge, skills, and attitudes (typically and) necessary for the performance of *core tasks* in our discipline, that is, *measuring* and *improvement*. The history of quality science has developed in the type of (knowledge) object, which, as in other disciplines, is done in stages. Usually beginning with: 1. Product and 2. Process. Opinions about Stage 3 differ and so even the names differ. Characteristic of the development is the increasing weight of the (measurement and improvement) problems which the quality experts

[1]Huub Vinkenburg is a quality care researcher and advisor. E-mail: huub.vinkenburg@hetnet.nl



T. Hardjono and E. van Kemenade, *The Emergence Paradigm in Quality Management*, https://doi.org/10.1007/978-3-030-58096-4_10

confronted in practice. The use of language is typical of the development. Quality scientists do not only speak about the *solution of tame problems*, but also about the *handling of tough issues.*

- *Quality management or quality care.* Depending on the context (things or people) the (my/our/his) discipline has a different name. For example: quality management or quality care (compare language use in the industry with language use in healthcare).
- *Quality object.* Some quality experts distinguish *three object types:* physical, psychic and social (It, I, We?), each with their own *category of (quality) criterion)*: resp. objective, subjective and intersubjective. Others prefer to see four categories, but then, what is the distinctive criterion?.
- *School.* The *school* concept refers to a group of people (i.e. *quality experts)*—fulfilling the core quality science tasks in "their" way, with "their" theories, models, instruments, and, above all, distinguish themselves from each other, especially "their" types of objects. (... *and in my opinion not by "their" paradigm?*).
  *Question:* does the term Goudse School refer to a group and/or to an idea?
  *And:* What next?

## 10.2 Thinking About Quality: A View (Kees de Vaal[2])

Long ago, I applied for a job as a quality manager. I was interviewed by a staff member of a bureau asking me at a particular moment whether I knew Pirsig's book *Zen and the Art of Motorcycle Maintenance.* Which was not the case. I immediately purchased the book and read it from cover to cover. Fantastic: an author who puts the concept of quality into perspective and looks at it from a philosophical perspective. His second book, *Lila,* was a bit more difficult for me to read, but it also contains noteworthy pieces.

A book in which Pirsig's work plays an important role—as in this book—can therefore count on my immediate sympathy. And now the book is written by two authors whom I have known for many years and of whom I have read most of the publications, that is even a greater pleasure. I think I can even reveal with some accuracy who wrote which section!

Sometimes, authors have asserted that quality management does not possess a foundation with extensive scientific foundation. It would be more of a practical skill or art. I do not agree. The modern roots of the discipline lie in statistics and that is carefully scientifically founded. Various models developed in the normative world have been extensively validated. And more and more well-founded theories are emerging, in which the use of concepts such as perspectives (Fisscher), paradigms (Vinkenburg), axioms (Van Kemenade), and algorithms (Van Schijndel) is not shunned.

[2]Kees de Vaal is lead-auditor, consultant and professor of quality management. E-mail: kees.de.vaal@gmail.com

The beauty of this book is that it rearranges concepts developed, described, and applied in recent decades. I am thinking of the quality management revolutions of Jouslin de Noray, the work of Shiba, the four-phase model of Hardjono, the schools of Vinkenburg, and the value systems of Beck and Cowan. Both authors searched for a fourth school that Hardjono called the pragmatic and Van Kemenade the contextual school. Thankfully, these hypothetical schools are now "dissolved". The intellectual effort has contributed to the creation of this book.

The authors assume that—I am paraphrasing—"all good things consist of four". This is logical to them, because they want to organize, to structure, and the number four standing for stability and structure in numerology fits well with their thoughts. Huub Vinkenburg is right when he says that three schools are enough: Three is the number of growth and the movement. Remarkable is the fact that the number five plays a role in this book. Everard Van Kemenade used it in his value orientations and Teun Hardjono applies it in the five phases of the INK management model (which, by the way, was derived from the Berenschot generation model developed by him that even had six levels). In numerology, the number five is the number attributing meaning, the number giving direction and meaning to the number four's structure. Five is an important number in world history. In China it is a lucky number, the linking number uniting Yin and Yang. The Old Testament includes the five books of Moses, the Pentateuch. The Buddhistic Dhamma (doctrine) contains five rules for life. Calvinism has five doctrines. Islam is based on five pillars. If we want to link numbers to the paradigms mentioned here, the number two (duality) belongs to empiric ("right or wrong"), three to emergence (movement, pluralism), four to reference (norms, structures), and five to reflection (connecting, meaning). Would there be a fifth paradigm, corresponding with number one (unique, new beginning)?

Teun Hardjono and Everard Van Kemenade call their new concepts *paradigms* and defend their use. I do not share this opinion. To me, paradigms have the meaning of a set of views, a frame of reference fitting a certain period of time with a paradigm shift occurring if the developments give cause to do so. Paradigms are largely mutually exclusive and follow one another. Paradigms have committed followers who fight each other. Examples are the control and involvement paradigms of Huub Vinkenburg and the Rhineland and Anglo-Saxon paradigms as elaborated by Jaap Peters and others.

The disappearing notion of *school* is less of a problem to me. I felt like a supporter of different schools which causes a bit of a problem: an artist belongs to a certain school and not a bit to one and a bit to another. The term dimensions is also in use (Quality in three dimensions, Ben van Schijndel). As far as I'm concerned, the concept of perspective comes closer, but in our field it is already linked to the systemic and socio-dynamic perspectives as introduced by Olaf Fisscher. Where there are actually four ways of orientation that may (and should) occur simultaneously—fitting in the context—and complementing each other, by being complementary. I argue for the use of the term "*view*" instead of "paradigm".

## 10.3 Unfolding Quality (Prof. dr. ir. Henk J. de Vries[3])

In his inaugural lecture, Hardjono (1999) compared attempts at defining quality to catching *mumsels*. They do not exist, but you still try to catch them. This book shows that the theme still has his attention. Together with Van Kemenade, he makes an admirable attempt to obtain a grip on quality. Whether or not this attempt has been fully successful yet remains to be seen, as it may be all too easy to classify a colorful variety of things under the four paradigms. Below, I want to try to stand on their shoulders and understand more about quality.

Earlier I defined quality as fitness for purpose (de Vries, 1999). That term also appears in this book, but I saw it in a broader sense: aligning with how something is meant. Then quality is a multi-dimensional concept, in which the dimensions coincide with the fifteen distinctive aspects of Christian philosophy ranging from the numerical aspect to the aspect of faith. The aspects are not reducible to each other. Each aspect is a window to observe reality. This is exactly what Hardjono and Van Kemenade have to say about paradigms.

Another concept of Christian philosophy is *unfolding*. That which has been provided in creation may be unfolded by man. He extracts what is already present in principle, while a rich variation may be possible. This process of opening up is visible in the breadth of entities to which quality applies; i.e. products, services, processes, people, organizations, and society. And in the attention for more aspects—ranging from technological to (also) economic to (also) aesthetic to (also) ethical to (also) religious aspects.

So our understanding of reality is limited—"Now we are looking in a reflection of a hazy mirror" (1 Corinthians 13:12). Measuring, which is so common in quality management, helps to understand reality, but may also be misleading: that which is not measured is excluded. In that case, quality control and quality improvement may even be going in the opposite directions. Hardjono shows something similar in his four-phase model. The Christian tradition uses the concept *sin*—missing the target, deviating from the "fitness for purpose".

In what Hardjono and Van Kemenade refer to as the Empirical Paradigm, the emphasis is on the given reality and getting to know this by measurements, and by management according to PDCA. Measuring is fine, provided we see the relativity thereof. Humanity has the divine command to rule creation (Genesis 2). This is also fine, yet contains the risk of derailment. Man can be mangled in anonymous control systems; the *system world* may rule over the "living world". This may be avoided by taking into account different aspects in a balanced way (de Vries, 1999; De Vries & Haverkamp, 2015).

In the Reference Paradigm another party will determine the requirements to be met. In this case, there is unfolding respectively toward the social and legal aspects.

---

[3]Prof. dr. ir. Henk J. de Vries is Professor of Standardisation Management at the Rotterdam School of Management, Erasmus University, and Visiting Professor at the Faculty of Technology, Policy and Management at Delft University of Technology. E-mail: hvries@rsm.nl, https://www.rsm.nl/people/henk-de-vries/.

The Reflective Paradigm adds disclosure to the psychic aspect. The Emergence Paradigm reveals more aspects and that is confusing. My previous booklet shows which they are, and indeed, confusion is partially inescapable, not only due to the multi-sidedness, but also due to the reflection in the hazy mirror.

In short, the four paradigms described in the book in their entirety in and of themselves are an example of unfolding. Yet they are incomplete. There are too many aspects lumped together. It is better to base the distinction of a parallel disclosure of:

(1) Entities (from product to society),
(2) Types of "requirements" for those entities (related to one or more of all fifteen aspects),
(3) Establishment of such "requirements" (with fewer or more different parties),
(4) Status of such *requirements* (from voluntary to imposed),
(5) Manners of measuring of meeting the requirements,
(6) Forms of control,
(7) Forms of improvement,
(8) Models that may be helpful for this.

Moreover, there is a broadening in the geographical sense (from local to global), and related to this from mono to multi-cultural. This in turn evokes the question of the intrinsic, the authentic. Which again relates to disclosure, since it is multi-faceted, the opportunities embedded in creation may be disclosed in many different ways. Failed disclosure is sin, which requires forgiveness and restoration.

The quality expert is the professional who, given the organizational situation, its context, as well as the organization's direction, who knows in which way the quality of products, processes, persons, and the organization in its entirety may gain shape, and who guides the organization in this respect.

## 10.4 A Philosophical and Scientific School in Quality Science (Ben van Schijndel[4])

Let me get straight to the point. The book *Thinking About Quality in Four Paradigms* originated from viewing quality from a philosophical perspective. The proponents of this, in my opinion, belong to the *philosophical school of quality*. There is nothing wrong with the philosophical school. With its paradigms that allow us to view the world of quality in different ways, we are inspired to improve this world. And yet something is missing. The book does not have another undeniable vision of quality. In my opinion this is the experiencing and studying of quality from a scientific perspective, rooted and inspired by the physical sciences. The same physical sciences in which Thomas Kuhn in his *Structure of Scientific Revolutions* was looking for

[4]Dr. Ben van Schijndel is a retired senior quality policy staff member of HU Utrecht University of applied Sciences Utrecht. E-mail: vanschijndel.ben@gmail.com

paradigms and paradigm shifts. I refer to the scientific approach as the *scientific school of quality*.

In Chap. 2, "The classification of Thinking About Quality", I attempted in vain to find an approach that connected to the scientific school. Further, in Chap. 4: "On Value Orientations and Paradigms" a brief reference is made to my publication *Kwaliteit is mensenwerk; relationele kwaliteit als kwaliteitsfactor*; (Quality is made by humans; relational quality as a quality factor). (*Synaps 2007* nr. 23; pp. 7–11). In this publication, Gerard Berendsen and I introduce professional/content-based, organizational and relational quality. In the following 10 years and 10 publications and 1 book further, I developed these topics into a quality-science-based theory rooted in aspects of the physical sciences. It is a theory that demonstrably has links with practice by its applications to the individual and collective level of professional/content-based, organizational, and relational quality dimensions. From a philosophical perspective, the aspect of *Quality is human work* may be approached in different ways. I am thinking of virtues and the awareness of values. However, the true basis and thereby the explanation and predictability of the behavior of the *human factor in quality* is a part of the evolutionary and behavioral biology. This describes and declares as well how and why learning and behavioral changes occur at an individual and collective level. In other words, what happens during continuous improvement. With this knowledge it is relatively simple to discover the impediments. If this does not happen, we are able to do something about it. Not every quality manager also needs to become a behavioral biologist. Yet an understanding of the basic principles as described in my publications is necessary.

What was Thomas Kuhn searching for in his book: *The Structure of Scientific Revolutions* where paradigms are concerned? Kuhn is clear about which sciences are concerned: physics, chemistry, biology (the physical sciences). Kuhn was looking for the origins of paradigms and the replacement of old by new paradigms: the paradigm shift. He argues that this happens by revolutions and not by gradual changes. I take note of his point. For me it is now more important to view the characteristics of paradigms from natural sciences. Based on this knowledge, I want to study the implications for a scientific approach of paradigms from quality science.

The "real" sciences (Kuhn refers to this as "normal sciences") share that they have paradigms in their domain standing for tangible solutions to particular issues. After a closer look, I came to the following distinction:

- *Scientific Paradigm: a cohesive set theories and models.*

In my own words: explaining and describing scientific paradigms and you can apply these to perform analyses for solving problems/issues. They comply with the principle of complementarity: They are mutually exclusive and complement each other in relation to one other.

- *Sociological/psychological paradigm: a constellation of beliefs, values and practices shared by the citizens of a given society.*

In this sense they are identical to the philosophical paradigms presented in the book.

In my view, they indicate a direction of thinking and acting in order to achieve a quality goal or socially desirable goal. However, they do not offer explanations. This requires the scientific paradigms.

"Thinking about Quality in Four Paradigms" in the scientific school is quite different. Below, I summarize the description and characteristics. I have described the complete discourse in Synaps 39 under the title "Paradigms and Algorithms in Quality Management".

- 1. The scientific support builds on the findings and inspiration of physics, evolutionary and behavioral biology, and learning theory.
- 2. Paradigms have to meet the *complementarity principle*. This term has deliberately been borrowed from theoretical physics. It meets the MECE-criterion: mutually exclusive, collectively exhaustive.
- 3. There are three scientific paradigms with their own algorithm:

    1. The system technical and social dynamic perspective (Fisscher);
    2. The three quality dimensions: professional/content-based, organizational, and relational. (Van Schijndel);
    3. The four quality schools are in fact no schools, yet meet the conditions of the scientific paradigm: the empirical, normative, reflective, and pragmatic school (Vinkenburg/ Hardjono);

- 4. There is one paradigm that does not have an algorithm: Control and Involvement (Vinkenburg).

One more comment: The concepts philosophical and scientific school of quality are not yet included in the article published in Synaps 39. I reserved the introduction for the current contribution to the book *Thinking about Quality in Four Paradigms.*

After my first explorations of both schools of thought, I cannot escape the impression that they are indeed complementary.

## 10.5 Thinking About Quality (Arnold Roozendaal[5])

*Introduction*

In the book Perspectives of Quality, I wrote a contribution about the developments in the quality manager area. My arguments related to the job of a quality manager who thus becomes the Program Manager Organizational Transformation: a job defining transformation. As described in the book, the profession of quality management in recent years has gained momentum. Changes require the quality manager to be aware

[5]*Dr. ing. A.H. (Arnold) Roozendaal (1954) is Senior Management Consultant and Senior Teamleider at Tüv Nederland and Internal Auditor at the Canisius Wilhelmina Hospital in Nijmegen. A member of the DAQ (Dutch Academy of Quality), he is part of the NNK working group on personal certification for (lead) auditor according to the EOQ standard. E-mail: info@contextueelleiderschap.nl*

and to change accordingly. It is a challenge to demonstrate his added value now and in the future.

Considering the four paradigms as outlined in this book, the same may be said of the internal and external auditors who ultimately test the implemented quality management system within an organization. To them, it makes a difference from which paradigm things are being viewed.

Organizations want to be convinced of the added value that an audit provides. Therefore, it is important that the (external) audit program recognizes the company's own (complex) questions about issues and that the deployed frame of reference is meaningful. So the (external) audit program must not be limited to determining conformity to the current agreements but contribute to the organization's strategic course. In many cases a shift will occur from testing based on the Empirical Paradigm and the Reference Paradigm to the Emergence Paradigm. In the following table, I show the relationship between the different paradigms, the roles of the quality expert, as described in this book, and the role of the auditors.

| Paradigm | Role of quality expert | Role of (lead) auditor |
|---|---|---|
| Empirical Paradigm | Bean counters, accountants, independent people, internal auditors | Diagnosis based on data |
| Reference Paradigm | Administrators and managers intending to assess the organization of quality and who put their trust in a model | Gap analysis at system level. The reference model of the norm is at the center. System audit and compliance |
| Reflective Paradigm | Anyone who, together with others, is willing to critically examine (learn from) their own actions. And who is willing to fulfill a role as a coach, mediator, therapist, friend or ... jester | Diagnosis, soft controls |
| Emergence Paradigm | Everyone who is open to other visions and models, who exercises self-criticism and is willing to engage in a dialogue with others. Change Agent | It is not the object that is at the center, but the way in which the entity is studied. The change process is at the center: from current to nominal values. This demands specific research capabilities |

In this contribution I want to focus on the role of the (lead) auditor, who views matters from the Emergence Paradigm perspective. I focus on auditors who carry out first and second party audits, that is, internal audits and customer/supplier audits, nationally and/or internationally.

*Importance for testing based on the Emergence Paradigm for (lead) auditors*

The above-mentioned developments have not gone unnoticed by the NNK (Netherlands Network for Quality Management). By the introduction of a European certification standard (EOQ) for auditors (and quality managers), the NNK took the initiative for recognition and legitimization of (lead) auditors tying in with the above-mentioned development. NNK compiled a list of criteria requiring compliance by auditors. This list contains important points of attention required for testing based on the Emergence Paradigm and containing testing criteria from four angles.

1. The auditor's competencies (audit and research abilities and change management competencies)
2. The ability to act as the customer's *critical friend*
3. To be a sounding board for the board (or the management)
4. Personal competencies:

Being able to list the criteria is one issue, while making sure that people are able to comply with them is another. Obviously, it is up to the (lead) auditor himself to give substance to this based on his personal leadership. From my own experience as a Certiked senior team leader, I know that the following points make a positive contribution to the (lead) auditors' personal development and professionalism:

1. Taking responsibility.

Based on the principle that the (lead) auditor assumes his responsibility to continuously evolve, so he is able to anticipate the future. It relates to the permanent personal development as a (lead) auditor.

2. Making use of existing frameworks.

NNK makes use of the competency framework of the EOQ (lead) auditor. This framework provides clarity of the necessary competencies to be met. The advantage of making use of this framework is its international recognition.

3. The organization of the company's own feedback.

A professional, as well as the (lead) auditor, is expected to reflect in many ways and to check himself. It is not only about reflection on the content but also about the inner dialogue of what drives him. This may be tested, for example, by means of casuistry or inter-vision. It involves a constant process of continuous reflection, improvement, and application in practice.

4. Obtaining personal certification.

For more information, please refer to the NKK website:
http://www.nnk.nl/contents/persoonscertificering_2

*Finally*

It's not that suddenly anyone is able to test from a different paradigm. Many of the current (lead) auditors have been trained in the field of system audits, which often matches their personality structure. It takes courage, an explicit decision, and continuous development of (lead) auditors to test from the Emergence Paradigm and thus make a substantial contribution to the organization's strategic direction. In that sense, we can speak of a (lead) auditor's job defining transformation.

## 10.6 The Primacy of the Emergence Paradigm from the Perspective of Art and Musicology: From a Prescriptive Model to Unfolding Reality. (Rik Spann[6])

*The mind is like a parachute. It only works when it is open.*
Frank Zappa

From the perspective of the truly creative mind, subdividing reality in a number of distinct paradigms is a counter-intuitive activity. And of course, not everyone views reality from this perspective, nor should we only view reality from this perspective. The value of the proposed classification in this book is great, leading to a useful expansion and nuancing of the view that many hold until now. As seen by the truly creative, naturally often holistically perceptive spirit, reality cannot be subdivided, reduced to isolated elements, tight categories and domains that do not overlap one another. The complexity of the world is a given. A priori, apart from exponentially growing technological explosions. Still, artists do deal with the *paradigm* phenomenon in a passionate way. From my personal experience I can remember a film that I developed as a visual artist with a group of fellow artists in my Artless program (a spin-off of the Rietveld Academy) around the *paradigm shift* phenomenon. During an exhibition of the work in the Oude Kerk, the theme yielded interesting discussions, for example in the dialogue with artists from Spain participating in the project. Cultural differences, artistic diversity, paradigm shifts: delicious spiritual nutrition to consume with hearty snacks and tapas.

Considering reality from a different perspective is a rewarding subject for many artists. Yet what the *paradigm* phenomenon means to an artist is often something fundamentally different from what it means to art theorists, or to artists who distance themselves from their work in order to consider it and understand it in language. Creating art and talking about art activates other brain and language centers. An artist—now I am speaking from my own experience, others may have a different opinion—finds himself/herself in the area under observation from a paradigm.

In this way, there is only one relevant paradigm: the *emergence paradigm.* To me, this means: you create a work (drawing, painting, object, installation, performance, composition) by making choices within the moment itself. Choices, **presenting** (which emerge) from every step you take. "Planning" means (expressed in musical terms): "setting the stage", or preparing the stage for the activity by means of which the artist will relate to reality unfolding in the moment. Anything which you might call "empirical", "referential", or "reflective", is mentally transported to a dimension related to the creative process as a map relates to the area, as an image of a pipe to the pipe itself. *Ceci, c'est une pipe, naturellement.*

---

[6]Drs. Rik Spann is a musicologist, improvising musician, visual artist and an organization scientist. He connects management and leadership with the essence of creativity. He is d co-founder of DeGoudse School. E-mail: rikspann@gmail.com

From a theoretical perspective, with a creative "twist", you might say that you can look *at* all paradigms *from* all paradigms. From referential to all four, from reflective to all four, etc. You may regard reflective as "from another logical order" than the other three; as it were reflecting to empirical etc. (and by itself). In other words, the rules of the game of the holistic, systemic reality in which the artist's mind usually feels at home—often without the possibility of escape—impose a framework on the game in which a theoretical treatise becomes legitimate only in the form of a game in which language and meaning may, or even must be, juggled.

From a musicological point of view, you allow for a fine brainstorm on the assignment of musicians, musical styles, or musical techniques to different paradigms. Here, too, the above comments apply. Yet if you did intend to say something about it—with as disclaimer the reference to the idea that art *in general* can be seen primarily from emergence and only in a derivative sense from the other angles—then there are interesting reflections, as thinking exercises, that one might apply.

Viewed from the room to maneuver typical of the creative mind:

*Empirical*:

Ignoring the *blue notes* in classical music theory, due to their "unmeasurability".

*Referential:*

The composition and arrangement principle "Theme and variations": norm and the newly created derived reality.

*Reflective*:

Reflection on different performances of the same piece—the gymnopédies by Satie by Aldo Ciccolini and Reinbert de Leeuw, and what this variety produces in the reflection on it—is a grateful and intriguing subject in my musicology lectures.

*Emergent*:

The role of "jazz"—a single word for a multitude of styles and approaches—as a style of art that often unfolds in the moment, developed in improvisation. In different degrees—interpretation, decoration, variation, free improvisation. In a wealth that does not allow classification without violating the essence of this wealth. (These examples, this matter and what this means for the greater context will be discussed in more detail in the planned book *Emergent Organizing* by Teun Hardjono and Rik Spann).

Fortunately, a one-to-one classification is not necessary. Only from the paradigms based on prescriptive models would an argument "stand or fall" by the grace of an attribute such as "exactness". But since art and music flourish in the company of elements such as chance, uncertainty, and chaos, accuracy is not the criterion. It is not about a proper classification but about the beating heart. Also in analogy with organizational practice something emerges here, and an interesting parallel looms here. After all, business administration is not an exact science, but a social science. The skill of organizing in the living balance with the art of watching: taking up the challenge of sharpening and blurring the perspective on the living, creative process in the moment itself time and again, in an immeasurable and unpredictable rhythm. At the service of the essence, the dynamics, the quality of both organizing tones in a musical whole and people in a vital, viable, healthy evolving organization.

## 10.7 Arend Oosterhoorn[7]

First of all, my compliments for the work, the structure ensures that the earlier discussion about the schools in a natural way has come to a clear division making further development of the ideas perfectly possible.

There are a number of considerations that I would like to add. When we talk about *quality*, we are indeed referring to certain qualities of an entity. *Quality* in itself is nothing, it is always something of something. These entities are also known by now. Product/service (outcome of the process), process (primary, supportive, controlling), organization, chain and as final (as remarked in the INK model, unfortunately replaced) society. To me, the separate categorization of process and profession appears to be superfluous. The description of Stevens et al. (1999) demonstrates that the professional aspects are aspects of the process.

When we look at the *quality* concept from the four paradigms, it must be possible to say something about quality of the mentioned entities.

When we look at the *measuring is knowing* paradigm, this particularly focuses on the determination of the entities' properties. For products this was developed ages ago. Since we made a go/no go mold in 1823, and Eli Witney by the beginning of the century experienced so much trouble with parts' variations that he was unable to assemble a functioning musket. Based on these, we have started measuring the properties and setup quality control. The fact that we could also do this for services is still under discussion, even though Zeithaml et al. have mapped the quality dimensions of a service. For a while now, we have been able to measure processes as well as organizations. The most pregnant example is scoring an organization by means of a self-evaluation on the basis of the INK model. Measuring the quality aspects of chains is unknown to me, but it will definitely happen. Societies are scored in different ways. Just think of the economic statistics or countries' scores on a bribery scale.

It is no more than self-evident that products are measured based on norms, referred to as tolerances. The discussion, first about the normative school and now the Reference Paradigm, specifies the results with which the entity measurement must comply. These are always externally formulated requirements demanding human effort so that the entity under observation will meet these requirements.

From these two paradigms it is useful to talk about the Plan-Do-Check-Act cycle. There is something such as good and poor quality, something that meets the requirements and something that does not. When it is good, *fitness for use* could be translated

[7]Arend Oosterhoorn is an advisor/trainer Lean Six Sigma and quality management. E-mail: aoosterhoorn@oosterhoornadvies.nl

as measurable properties of the entities that must realize the fitness. It did not succeed for all entities.

Now, let us look at the emergence paradigm. Variation is also an emergence phenomenon. One single product (entity) does not contain any variation. With two or three items, there is not that much information available. You need more items to obtain a good overview of the pattern of variation. The question asked most frequently to the statistician is, therefore: "How many observations must I do to get a good view of ... ". We don't see variation until various (comparable) products (entities) are observed. This also applies to services, professionals, organizations, societies. Authors concluding that Deming (and Shewhart as his teacher) base their views on the emergence paradigm are no coincidence. He observed the world from the statistical point of view, with all aspects showing variation. In order to be able to steer, controlling the quality of the entities (human effort) you must understand the entire process. In order to understand problems and be able to solve them, the situation in which they arise must be studied. Such study is only possible when the proper methodologies are applied. Statistics and Six Sigma are no instruments of the Empirical Paradigm but of the Emergence Paradigm. This does not concern the PDCA cycle, but Plan-Do-Study-Learn. Even the approach in the Cynefin model, with complex matters Probe-Sense-Respond, is here actually the mechanism. Trying to see how the system responds to learn from this is an important, yet underdeveloped, quality-science principle.

All measurements have their limitations, as well as the fact that we can call into question all specifications and standards. And our knowledge of the entities that we, as quality experts, think we should say something about, is also limited and open to much discussion.

When you call everything into question, it appears to be a weakness, yet in fact it is a strength. It's just that *The more you learn, the more you realize, the less you know* or as Francis Bacon said: *If a man will begin with certainties, he shall end in doubts; but if he will be content to begin with doubts he shall end in certainties.*

It is also wise to understand the context. Time, place, development level of the entity are context-related aspects. Talking about *Quality* therefore means that you are talking, always with modesty, about the things that could best be done from a quality-science perspective (emergent phenomenon) at this moment and in this situation.

If we, as quality experts, recognize these mechanisms, we are also tasked to prevent disorder. Although, who knows the beautiful things that might result. We don't know.

Snowden in his Youtube film distinguishes in *categorization* and *sense making* models. This is also reflected in the paradigms. The Emergence Paradigm is the sense-making variant of the Empirical Paradigm, the Reflective Paradigm that of the Reference Paradigm. It gets really fun when Reference Paradigm hits the Empirie Paradigm. Snowden calls the separation between chaotic and simple a cliff. But thinking in the paradigm poses the essential challenge. What was the point again? Everything is called into question. What we perceive we have to interpret from our own references, and the earlier images are no longer sufficient. Do I understand what

I see and say? What meaning do I give to what I perceive? How do I interpret it and what I think of it (from my frame of reference)? This is about the question of what you think of it yourself, now that you perceive *reality* and are asked to make a (quality-wise) judgment about it. This is not based on (external) standards, but on your own reference. This initiates THE DISCUSSION in which people are going to talk about human work and its quality. Then people will come together, *reaching the essence.*

## 10.8 Olaf Fisscher

**Integrity as the essence of quality**

Within the framework of the post academic Master's Degree program Risk Management on the study trip to Stavanger, visiting Statoil, the Head Risk Management in closing gave wise advice to the students in the form of three short statements: 1. Know the facts! 2. Don't only think of negative risks, think also of positive risks! 3. Have integrity!

To me it's all about the emphasis on *integrity*, surprising in an environment that seems to be exclusively focused on control... or maybe not so surprising perhaps?

That much is clear: the normative-ethical dimension is on the map, for risk management as well as for the most closely related quality management. It is time to give this achievement the necessary recognition of a choice of a separate school or a separate paradigm. It seems most appropriate to me in this respect to focus on the *integrity* concept. In Vinkenburg's order this could be referred to as a fourth school, the *Integrity school*, or in Hardjono's order as a fifth paradigm, the *integrity paradigm.* The latter sounds better. Therefore, it has my preference.

***Distinguished schools and paradigms***

Vinkenburg and Hardjono talk about three, respectively, four distinguishable approaches within the quality domain. Such a classification is useful and clarifying, both in the context of historiography (in so far as the classification represents a development) and in the context of specification of the concept and meaning of *quality.* As I understand it, both authors claim to be offering a combination of both. The question then is whether each subsequent approach includes the previous one, whether approaches are mutually exclusive, or whether they can coexist. This is an interesting question within the framework of developing a model. However, I do not yet see a clear answer.

It seems to me that quality science is an applied science requiring not only specific knowledge but also skills and experience, as well as context-specific professional action on the part of the quality expert. In these applied scientific knowledge domains, it is appropriate to think in terms of *best practice* and *evidence-based practice.* Hardjono and Van Kemenade discuss this in a very clear fashion for health care quality.

Unlike in the natural sciences, where nature itself has the last word, in the field of quality science, what *application practice teaches us* also applies.

The attractive thing about speaking of "schools" is that we also differentiate between categories of professionals grouped around a particular view and the associated knowledge and experience base, thus forming a school together, such as the Goudse School by Hardjono and Roozendaal, for example. Schools are also the places where professionals are educated and socialized within the specific professional group. They may also be locked in as a one-sided trap. Specific knowledge and craftsmanship on the one hand form the basis and strength of the professional in question, but on the other hand also form a potential limitation, which arises if knowledge and skills are not kept up to date, or if they are not trained any further or retrained elsewhere.

The distinction made by Vinkenburg in an empiric, normative, and reflective school appeals to me. I find Hardjono's expansion of the three paradigms he distinguished earlier quite exciting and useful. With this distinction he challenges us to improve our understanding of the typology of approaches in quality-science. I would want to challenge Hardjono with respect for the sanctity of the number four, to also leave room for other numbers, invented for a reason. Why is there no fifth paradigm or *worse*? Hardjono by further exploring his extensive, in-depth knowledge of the history of quality management (a history that he himself co-authored) can undoubtedly still delight us with a multitude of additional insights. Models are sometimes merely useful tools. Is not the same message typical of the fourth, the *Reflective Paradigm*? This latter paradigm can hardly serve as a reservoir for everything yet to come under the influence of the context. In my view this reveals an internal inconsistency.

In addition to his third school (the *reflective school*), Vinkenburg argues in favor of having an eye for the ethics of virtue. I want to agree. In my opinion, the ethical norm of *integrity* is at its best if we base ourselves on virtue as an ethical theory, and not exclusively on ethical theories such as *duty ethics* and *consequential ethics*. Integrity stands for wisdom, weighing situations and possibilities with one's own conscience, making choices based on one's own values and being prepared to justify them.

In short, I propose a new paradigm: the *integrity paradigm*, in line with Vinkenburg's plea for virtue as an ethical theory, embedded in the concept of *integrity*.

By discussing the *integrity paradigm*, we explicitly and substantively address the fact that this concerns the normative ethical dimension. To me, treating the appeal to everyone's conscience by neutral, specific labels of insufficient meaning, such as *reflective* and *emergent* is undesirable since it shrouds the essence.

### *Integrity: a beckoning, winning perspective*

Integrity may be described as: acting based on a sound judgment and on searching for a unity or wholeness of thinking and acting. Essential for *integrity* is the combination of soundness and wholeness/ cohesion. Values and behavior have to match one another. Good intentions are not sufficient. This means, for example, that an organization's effective code of behavior as well as its activities have to match in order to be convincing.

Characteristic for *integrity* entails that it focuses on oneself: the key elements of this are self-awareness, conscience, and inner transparency.

Even with organizations we may conclude that their approach is convincing, that the products and rendered services are valuable to society, and that the intentions and expectations are met. Quality management provides an essential contribution to the organizational professionalism. It is not enough for professional organizations to only provide a number of sufficiently qualified employees. They also need to provide an organization which as such may be called professional. Integrity stands for connection, cohesion based on professionalism and responsible behavior. This moral dimension is the human foundation of organizations. Based on a climate of impeccability and of virtuous action, the passion and ownership of one's own task and of the organization as a whole will come into being.

***An organization with integrity***

Employees and management together carry the organization, bring it to life and produce the result. For everyone in the organization, and also for the organization as such, the principle is that they must give shape to how they function based on their own values. It is a case of reciprocity, provided that the one is not hiding behind the other. It is a moral, normative task to want and be able to represent an organization. Based on a personal integrity, this results in a shared collective integrity, a wholeness, and value-driven approach as the essence of quality.

Olaf Fisscher, Emeritus Professor in Quality Management and Business Ethics. E-mail: o.a.m.fisscher@utwente.nl

## References

Hardjono, T. W. (1999). *Laveren tussen rekkelen en preciezen*, Oration Faculty of Business Administration at the Erasmus University Rotterdam. ISBN 90-56773127.

Stevens, F., Bering, R., & Stevens, K. (1999). *Leren excelleren* 12 Jaar General electric tegen de achtergrond van het INK-model.

de Vries, H. J. (1999). *Kwalitetszorg zonder onbehagen*. Amsterdam: Buijten & Schipperheijn.

de Vries, H. J., & Haverkamp, A. (2015). Overcoming resistance against quality control—A philosophical-empirical approach. *International Journal of Quality and Reliability Management, 32*(1), 18–41.

## *Uncited Reference*

Burrell, G., & Morgan, G. (1979). *Sociological Paradigms and Organisational Analysis: Elements of the Sociology of Corporate Life*. Aldershot: Grower. ISBN 0566051494.

Hardjono, T. W., & Beusmans, H. P. G. *Europa Heruitvinden*. Deventer. Vakmedianet. ISBN 9789462761421.

Kanji, G. K., & Asher, M. (1996). *100 Methods for total quality management*. London: Sage Publications. ISBN 0-8039-7747-6.

Kemenade E.van, Pupius M. en Hardjono T. (2008). More value to defining quality. *Quality in Higher Education*. Retrieved 4th of March from https://www.researchgate.net/publication/225083599_More_Value_to_Defining_Quality.

van Marrewijk, M., & Hardjono, T. W. (2003). European corporate sustainability framework for managing complexity and corporate transformation. *Journal of Business Ethics, 44*(2–3), 121–132.

Mintzberg, H. (1979). *The structuring of organizations: A synthesis of the research*. University of Illinois at Urbana–Champaign's Academy for Entrepreneurial Leadership Historical Research Reference in Entrepreneurship.

Mintzberg, H. (1994). The fall and rise of strategic planning. *Harvard Business Review, 72*(1), 107–114.

Plato. (1986). *Protagoras*. Translated by B.A.F. Hubbard and E.S. Karnofsky, in The Dialogues of Plato. New York: Bantam Books.

Schön, D. A. (1987). *Educating the reflective practitioner, toward a new design for teaching and learning in professions*. San Francisco/London: Jossey-Bass Publ.

Stacey, R. (1996). *Complexity and creativity in organizations*. San Francisco: Berrett-Koehler.

Stacey, R. (2011). *Strategic mmanagement and organisational dynamics. The Challenge of Complexity*. London: Pearson.

Weick, K. E., & Westley, F. (2002). Organizational learning: Affriming an oxymoron. In S. R. Clegg, C. Hardy, & W. S. Nord (Eds.), *Handbook of organization studies* (pp. 440–458). London: Sage.

MIX
Papier aus verantwortungsvollen Quellen
Paper from responsible sources
FSC® C105338

If you have any concerns about our products,
you can contact us on
**ProductSafety@springernature.com**

In case Publisher is established outside the EU,
the EU authorized representative is:
**Springer Nature Customer Service Center GmbH**
**Europaplatz 3, 69115 Heidelberg, Germany**

Printed by Libri Plureos GmbH
in Hamburg, Germany